LOS BRAZOS DE DIOS

Conflicting Worlds

NEW DIMENSIONS OF THE AMERICAN CIVIL WAR

T. Michael Parrish, Series Editor

LOS BRAZOS DE DIOS

A Plantation Society in the Texas Borderlands 1821–1865

SEAN M. KELLEY

LOUISIANA STATE UNIVERSITY PRESS
BATON ROUGE

Published by Louisiana State University Press

Manufactured in the United States of America
First printing

DESIGNER: Mandy McDonald Scallan
TYPEFACES: text, Whitman; display, No. 13 Type, No. 12 Type, and No. 11 Type
PRINTER: McNaughton & Gunn, Inc.
BINDER: John H. Dekker & Sons

Library of Congress Cataloging-in-Publication Data

Kelley, Sean M., 1966–
Los brazos de Dios : a plantation society in the Texas borderlands, 1821–1865 / Sean M. Kelley.
p. cm. — (New dimensions of the American Civil War)
Includes bibliographical references and index.
ISBN 978-0-8071-3687-4 (cloth : alk. paper) 1. Brazos River Valley (Tex.)—History—19th century. 2. Brazos River Valley (Tex.)—Social conditions—19th century. 3. Plantations—Texas—Brazos River Valley—History—19th century. 4. Plantation life—Texas—Brazos River Valley—History—19th century. 5. Slavery—Texas—Brazos River Valley—History—19th century. 6. Immigrants—Texas—Brazos River Valley—History—19th century. 7. Brazos River Valley (Tex.)—Emigration and immigration—History—19th century. 8. Texas—History—Civil War, 1861–1865. I. Title.
F392.B842K455 2010
976.4—dc22

2010006408

The paper in this book meets the guidelines for permanence and durability of the Committee on Production Guidelines for Book Longevity of the Council on Library Resources. ∞

To Vicki

Contents

Acknowledgments

This study has had a long and varied life, starting as a seminar paper, growing into a dissertation, and finally reaching maturity as a book. A great many people and institutions have helped over the years. Fellowships from the University of Texas at Austin, including three Schumacher-Adkins Fellowships, helped fund dissertation research. A Hartwick College Faculty Trustee Grant allowed me to return to Texas for research in the summer of 2002. Archivists at the Dolph Briscoe Center for American History at the University of Texas at Austin, the Rosenberg Library, the Hill Memorial Library at Louisiana State University, the Southern History Collection at the University of North Carolina, the Baker Library at Harvard Business School, the Earl K. Long Library at the University of New Orleans, and the National Archives were all indispensable in helping me find scattered sources. The district and county clerks of Austin, Brazoria, and Washington counties graciously allowed me access to their holdings. Marie Meyer, district clerk of Austin County, and her staff deserve special thanks for allowing me the run of the vault and never complaining when I got in the way. Jamie Murray, at the Brazoria County Historical Museum, also deserves a special word of gratitude for going out of her way to share the museum's many valuable resources and filling me in on the intricacies of local history and geography. Galen Greaser, at the General Land Office, not only helped me locate important materials; he straightened me out on several confusing points of nineteenth-century land policy. Ed Bentley, of the Chappell Hill Historical Society, allowed me access to that organization's considerable collections and introduced me to Nath and Judy Winfield, whose knowledge of the region I could never hope to match. Sally van Dyke, of the Brenham Heritage Society, donated her time to show me the Giddings-Wilkins house and allowed me to examine transcripts from a collection still in private hands. Ginny Raska very generously gave over her dining room table for the better part of a week so that I could examine the Levi Jordan Papers.

Many people provided feedback on various aspects of the project, at conferences and seminars and in private. My editor, Michael Parrish, saw the potential of this project and improved it greatly with his suggestions. Bill Ashbaugh, Roark Atkinson, Thomas Beal, Nancy Bercaw, Bill Blair, Anne Boylan, James Brophy, Randolph B. Campbell, Gregg Cantrell, Matt Childs, Lisle Dalton, Dora Dumont, Judith Fossett, Richard Graham, Amy Greenberg, Matthew Hendley, Watson Jennison, Kevin Kenny, Robin Kilson, Cherilyn Lacy, Hal Langfur, Brian Larkin, Sandra Lauderdale, Marc McLeod, Eric Meeks, Randall Miller, Todd Moye, and Catherine Nolan-Ferrell all took time to offer criticisms and suggestions at different stages. Michelle Lerner hand-drew the maps. The late Terry Jordan-Bychkov served on my dissertation committee and shared his vast knowledge of German Texas with me. Neil Foley first suggested that my seminar paper had "legs" and served as a dissertation reader. Jim Sidbury had many helpful comments; it has been a special pleasure to continue our conversations in recent years. Bob Olwell deserves special mention for reading innumerable drafts over the years, a deed above and beyond the call of duty. Dave Bowman not only directed the dissertation; he taught me what it was to be a historian. Peter Kolchin read and commented in detail on the entire manuscript and has proven a most gracious friend ever since my time in Philadelphia. Finally, I would like to give special thanks to Walter Kamphoefner for sharing his sources and insights, reading my work, and providing unfailing encouragement and support over the years. It goes without saying that any and all errors are entirely my own.

This book would not exist, and indeed I would probably not be in this profession, if it were not for Vicki Howard. Thank you. And thank you, Kathleen. Your arrival as I was finishing the manuscript forced me to take many a welcome break.

LOS BRAZOS DE DIOS

Introduction

When the young Ohioan Rutherford B. Hayes visited his college friend Guy M. Bryan on the Lower Brazos in the winter of 1848, he was impressed with what he saw. His lodgings at Peach Point, the plantation owned by Bryan's stepfather, James F. Perry, were "delightfully situated in the edge of the timber, looking out upon a plain on the south extending five or eight miles to the Gulf." The residents and neighbors he found to be "exceedingly agreeable, gay, and polished," and in the weeks he spent there, Hayes embraced the ways of the local elite, hunting deer and panthers, flirting with belles, dancing until dawn, and playing chess. The food he found especially good and plentiful, one meal consisting of "seven or eight kinds of meat, sweet potatoes in two or three shapes, half a dozen kinds of preserves, and pastry in any quantity." Of the region's slaves, who made up a majority of the population, Hayes saw "nothing to change my Northern opinions." Yet more than he was willing to admit, his experience seems to have disposed him favorably to the planters' view of slavery. "We have seen none of 'the horrors' so often described," he wrote to his mother, and he even suggested to her that the "Sambos" tried their masters more than the masters imposed on their slaves. Other than to note the isolation and the "wildness" of the country, Hayes sketched what he took to be a typical southern plantation.[1]

There was much, however, that escaped Hayes's attention, for Peach Point was not quite a typical southern plantation. For one thing, Hayes apparently never noticed that at least two of the slaves at Peach Point, and hundreds more on the neighboring plantations, were African-born. At a time when perhaps 2 percent of the black population of the United States was African-born, between 15 and 30 percent of the blacks in Brazoria County, the site of Peach Point, had experienced enslavement and the Middle Passage.[2] Hayes also missed the region's German immigrant population. Although the center of the Brazos German population lay two counties upriver, there was a small enclave in the town of Brazoria, not far from Peach Point. In fact, Perry did business with a German merchant there and occasionally hired Germans to do work on his plantation, such as fixing his sugar mill. Lastly, though he visited less than a year after the Treaty of Guadalupe Hidalgo, Hayes had nothing to say about the plantation's origins under Mexican sovereignty. Here, Hayes's oversight is more understandable—Mexican rule was more than twelve years past, and its influence was now subtle. Still, on several occasions Perry employed a Mexican teamster to move produce between Peach Point and his stock farm a few miles away.[3]

This book is about a slave society, one that was in many ways "typical" of the antebellum southern United States but at the same time quite different. Historian Ira Berlin's call, now more than a quarter century old, to pay greater attention to variations in "time and space" in the history of slavery has transformed the field. The tendency of historians to treat North American slavery as a monolithic institution transcending temporal and geographic variation has subsided in recent years, replaced with a far more nuanced and varied perspective. Virtually all would agree that the eighteenth-century Carolina Low Country was a world apart from Chesapeake, Louisiana, or New York and that antebellum Maryland was quite different from Florida or Mississippi.[4]

In keeping with this geographic and temporal sensitivity, the primary assumption of this book is that antebellum plantation society had a spatial dimension and that there is considerable value in examining slavery in the local context. Every place was not exactly like every other place, and different localities developed in different ways, at different rates, and in different political contexts. Rather than seeking out the typical, this book unapologetically emphasizes the unique and the exceptional. No doubt some will see this as an intellectually anarchic position, and to them I respond that the "typical" seldom exists and that all places and things are unique in some sense. Moreover, exploring the exceptional is valuable because it allows us to understand the limits of the norm, as well as to appreciate the immense variability of historical experience.

Texas was the only slave-based plantation society to originate under Mexican sovereignty, a circumstance that shaped its development profoundly. Although the demographic profile of the lower Brazos resembled that of Mississippi or Louisiana more than it resembled the classic southwestern borderland societies of California, New Mexico, and South Texas, the influence of Spain and Mexico was palpable. Mexican colonization policy deeply influenced the demographic structure of the Brazos, its land distribution, its marriage and property laws, and its labor relations, particularly the master-slave relationship. Moreover, these influences continued to be felt through the antebellum period. While the lower Brazos counties grew dramatically in the twenty-five years following political independence in 1836, and though they were formally annexed to the United States in 1845, their fundamental legal and social structures remained the same as the ones that had formed under Mexican sovereignty.

Lower Brazos slave society was the product of several distinctly nineteenth-century historical developments. The first of these was the hemisphere-wide expansion of plantation agriculture, which was driven by a growing consumer demand for slave-grown products in industrial Europe and the northern United States. The products included Cuban sugar, Brazilian coffee, and (most relevant for the Brazos) southern cotton. The second development, the emergence of an international abolitionist movement, seems ironic in light of the first and is the subject of a rich literature on the relationship between slavery and capitalism. It is important here not only because northern abolitionism affected the Brazos through the Texas annexation controversy and the Civil War, but also because abolitionism caught hold in Mexico, much to the consternation of Brazos slaveholders. The third influential historical development was the Latin American independence movement, beginning with Mexico, which freed itself from Spanish rule over the years 1810–1821. All of the newly independent American states inscribed their sovereignty on the land through the drawing of boundaries. In Mexico's case, it accepted an 1819 boundary with the United States that was worked out by Spanish negotiators and ran along the Sabine, Red, and Arkansas rivers, then along the forty-second parallel out to the Pacific. Lastly, the expansion of the capitalist market into new areas, such as central Europe, disrupted older patterns of agriculture and labor, spawning a wave of out-migration. Taken together, the expansion of slavery, the emergence of an abolitionist movement, the emergence of an independent and bounded Mexican republic, and a wave of emigration from the German-speaking areas of central Europe help account for the many contours of Brazos slave society.[5]

Although it seems quite obvious and natural, the advent of a bordered continent was a relatively recent development that entailed significant qualitative change.[6] In pre-Columbian America, and even during much of the Spanish period, the continent was borderless, the site of myriad interactions between different groups. Native Americans tended to conceive of space in terms of usage, making it possible for two or more groups to share the same land. Spain itself saw little reason to draw boundaries as long as its monopoly on continental dominion remained unchallenged. With the coming of rival claimants in the seventeenth century, the colonial powers turned to formal boundaries as a way to order their relations. Yet even these borders were more fluid than might be expected. Maps from the period often depict different colonial boundaries, depending on whether the map was English, French, or Spanish. These differences, often at variance with geopolitical reality, underscore the provisional, even wishful nature of most imperial claims. Indeed, many imperial claims were openly disputed or defied. The trans-Allegheny rivalry between Britain and France, which culminated in the French and Indian War, as well as the trans-Mississippi rivalry between France and Spain, which ultimately provided the impetus for the Spanish settlement of Texas, are just two examples of this fluidity.[7]

As historians Jeremy Adelman and Stephen Aron point out, the residents of these areas, the indigenous and mixed-ancestry "people in-between," maintained their autonomy by playing imperial powers off one another. As the former colonies became independent states in the late eighteenth and early nineteenth centuries, they negotiated ostensibly permanent territorial boundaries (though technically, the Adams-Onís Transcontinental Treaty of 1819, which set the southwestern border of the United States, was negotiated in the waning years of Spanish rule). Eventually, the imperial "borderlands" became "bordered lands," or independent states whose sovereignty was circumscribed spatially within internationally acknowledged boundaries. Because their political systems were rooted in the notion of "the citizen" rather than the monarchical subject, the new bordered republics tended to be more exclusive than their predecessors. In addition, noncitizens "in-between" generally experienced diminished autonomy now that the sovereign government held exclusive negotiating rights.[8]

The simultaneous geographic expansion and political rejection of slavery, along with the transition from borderlands to "bordered lands," frame our story. If borderlands are, according to Adelman and Aron, "the contested boundaries between colonial domains," distinct from the more static, less contested bordered lands, the history of the Brazos involves the transition from the former to the latter. That transition, however, was messy. Lower Brazos slave society took root in a borderland that was disputed by two sovereign republics, one in which slavery was ascendant, and another in which it was moribund. For a decade and a half, as slavery took hold, it existed in a statutorily gray area—a legal and constitutional borderland. After 1836, when political independence put the institution on firm legal footing, the newly drawn (and still disputed) border beckoned to enslaved Texans, making the institution chronically unstable. Even after the annexation of Texas by the United States and the decisive military defeat of Mexico, slavery persisted as one of the things that defined the Texas-Mexico border, one of several ways in which the border was made palpable. More subtly, the legacy of Spanish and Mexican rule influenced patterns of landholding, migration and demography, capital accumulation, marital relations, and master-slave negotiations. The Brazos of the antebellum period, then, was a particular variant of the southern slave society that spawned it: it was a borderland plantation society.

Historians have generally not seen the Brazos or other Texas plantation regions as a part of the U.S.-Mexico borderlands. More often, they have viewed them as unalloyed recastings of Anglo and African American slave society. They have sound reasons for emphasizing that point. In recent decades, many white Texans have clung to the myth that, though Texas was a slave state and part of the Confederacy, slavery had somehow been a minor and inconsequential institution. Seeking to dispel the myth, historian Randolph Campbell argued in his definitive treatment of slavery in the Lone Star State, "So long as Texas is not seen as a southern state, its people do not have to face the great moral evil of slavery and the bitter heritage of black-white relations that followed the defeat of the Confederacy in 1865." Noting that the proportion of slaves in the Texas population in 1860 was about the same as in Virginia, he suggested that "slavery was as strongly established in Texas, the newest slave state, as it was in the oldest slave state in the Union." To stress the distinctiveness of Texas slavery would have blunted the force of his argument and put off the historical reckoning he hoped to inspire.[9]

Other historians have proposed that Texas slavery was distinct, in that it was of the "frontier" variety. But many of these historians, in a laudable effort to take seriously variations in time and space, seized upon the notion of "frontier slavery." Few, however, bothered to define it, and worse yet, many invested it with Turnerian characteristics that do not reflect the perceptions or lived experience of the enslaved. Frederick Jackson Turner's now familiar argument, published in the 1890s, posited "an area of free land" that had shaped American culture by leveling Old World hierarchies. Turner's frontier took the form of a line separating civilization and savagery, and American history was essentially the process of pushing the line ever westward. As older seaboard areas stultified and grew more like their Old World progenitors, the westering process kept the nation vibrant, individualistic, democratic, and (most problematically for slave studies) egalitarian. Critiques of Turner's thesis appeared almost immediately, but his formulation remained influential into the 1970s. Early critics pointed out that Turner's definition of frontier was hazy—was it a line or a place? They also noted that his "thesis" was in fact a complex of interrelated subtheses involving the origins of democracy, individualism, labor, the economy, the family, and the innate capabilities of the races. Later critics, influenced by the Chicano and American Indian movements, as well as by the anticolonial impulses of the 1960s, elaborated a now-familiar critique of Turner's frontier as ethnocentric in its focus on Anglo Americans, imperialist in its imaginings of an "empty" continent full of "free" land, and racist in its willingness to discount the humanity of nonwhite residents. Where the term *frontier* survives, it has been completely redefined and stripped of its Turnerian implications. David Weber has recast frontiers as "zones of interaction between two different cultures"; another group of historians speaks of frontiers as "zones of fluid, ongoing conflict." Each definition has more in common with the current understanding of borderlands than with Turner's frontier.[10]

None of this is to suggest that those historians who embraced the idea of frontier slavery harbor the same racist and imperialist notions as the early Turnerians. However, in taking the concept of a frontier as applied to slavery for granted, they have taken on all of the messiness associated with it. It has been used to describe such diverse situations as early Carolina, the Old Southwest, and the sertão of Brazil. This does not mean that slavery in the interior was the same as slavery in coastal plantation societies; on the contrary, a single term cannot possibly address the range of conditions and experiences in all of those places. More importantly, in the absence of an explicit definition, the invocation of "frontier slavery" and its close cognates, such as "pioneer plantations," assumes certain conditions to be understood. Since *frontier* has often not had a definition so much as a complex of associations, there is an open question regarding which of these was the most influential for slavery.

Was it the availability of "free" and "empty" land that necessitated the use of unfree labor in the first place? The sparseness of settlement that resulted in the mutual dependence of master and slave? The distance from the seaboard and the leveling of its hierarchies? A labor regime centering on ranching and lumber rather than agriculture? The presence of native and other groups? Or was it the predominance of marginal slaveholders who were more inclined than their seaboard counterparts to extract labor through the use of the lash? A reasonable reading of recent scholarship says that each of these factors played some role, but their grouping under the rubric of *frontier* obscures their relative importance.[11]

If historians of slavery have frequently overlooked Texas's border situation, borderlands historians have largely ignored Texas plantation society, except to note its role as the hearth of Anglo American rebellion against Mexico. Again, the reasons are understandable: Central and East Texas looked too much like the southern United States, too black-and-white, with only a miniscule Hispanic population. The lower Brazos is a perfect example of this pattern, with an African American majority of 55 percent in 1860 (see table 1) and a very small Hispanic and Indian population (James F. Perry's occasional use of Mexican teamsters notwithstanding). With a demographic structure like that, many of the common themes of borderland history seem a poor fit: almost no Hispanic population, a small Indian population that was quickly expelled, and consequently, no mestizaje (or at least none among people of European and Native American descent). Add to that the dearth of missions, the prevalence of Protestantism, and the dominance of English, and the Brazos seems drastically different from Santa Fe or even San Antonio.[12]

However, the fact remains that the Brazos plantations were founded under Mexican sovereignty, and if we dig slightly deeper, the borderland influence becomes apparent. If one of the themes of borderland history is demographic diversity, the Brazos falls squarely in the mainstream. Of the five thousand slaves in the region in 1840, perhaps as many as one thousand had been born in Africa. Moreover, their presence can be explained by the region's border situation: slave traders took advantage of the ongoing conflict between Mexico and Anglo American residents to smuggle the Africans in from Cuba. The white population was even more ethnically diverse. Approximately 15 percent of Brazos whites came from the German-speaking areas of central Europe. If we count their Texas-born children, German-speakers accounted for about

30 percent of the free population on the eve of the Civil War. The original German settlements on the Brazos were founded in the early 1830s, a time when the Mexican government was growing increasingly wary of Anglo American immigration. During the 1850s, as the sectional crisis intensified, many German residents proved insufficiently supportive of slavery and "southern rights." Brazos German identity came to be seen by all parties—Anglos, African Americans, and the immigrants themselves—as an oppositional one. Tension between Germans and Anglos persisted into the postwar period, but the settlements continued to attract immigrants until the end of the century. The demography of the lower Brazos, then, was profoundly shaped by its situation in the Mexican borderlands, on both sides of the color line. It was a borderland unto itself.

The Brazos was a borderland society in still another sense, in that Mexico loomed large in the minds of its residents. It is relatively easy to demonstrate that Mexican settlement policy was responsible for the basic demographic mix and that Spanish-Mexican law influenced marriage and property rights. It is more difficult to demonstrate the importance of Mexico in Brazos residents' consciousness, but it was undoubtedly great, particularly among the enslaved, who viewed the republic to the south as a symbol of freedom. In the years before the Civil War, thousands of Texas slaves fled southward across the Rio Grande. The instability and economic costs of such activity were significant enough for the Texas Rangers to mount a recapturing expedition in 1855, which, of course, violated Mexican sovereignty. All of this suggests that, despite the lines on the map, Mexico figured significantly in the minds of Brazos residents of all colors and conditions.[13]

If we accept the notion that the antebellum Brazos was, at least to some degree, a borderland society, we must also recognize that the relationship between Brazos plantation society and the Texas-Mexico border changed over time. Historians of the American South and emancipation are accustomed to thinking in terms of change and continuity. As a slave society, the Brazos experienced all of the changes associated with the post–Civil War shift to free labor. But the destruction of slavery accomplished something more for the Brazos: it completed the transition from contested borderland to more static "bordered land." As long as Texas remained a slave state and Mexico a free republic, the pull of the border was strong. After 1865, slave flight and masters' attempts at recapture ended. Former slaves were far less likely to undertake the journey southward, although a few did. Most chose instead to stay in Texas and participate in the making of a new order. With slavery gone, Mexico's significance receded in Brazos residents' consciousness. Needless to say, the declining influence of the border on Brazos society did not mean that Texas as a whole went through the same transformation. South Texas remained (and most emphatically remains today) the consummate border society. The history of the Brazos merely suggests that the borderlands are as much a product of consciousness as of geography. The truth of that statement is borne out by the history of the Brazos in the twentieth century, as NAFTA and a resurgent Mexican population have helped turned the attention of the region southward once again.

Such an atypical place requires the use of terminology not usually seen in works of southern history. Specifically, I have adopted the term *Anglo* to refer to whites who spoke English as their native language. Use of *Anglo* as a proper noun is conventional in border studies, and it seemed preferable to *Americans* for a discussion of the Brazos, given the region's large German population and the simple fact that Mexican-born inhabitants are also Americans. I am also aware that the word *Anglo* carries no inherent racial connotations and do not mean to suggest that *black* and *Anglo* are mutually exclusive. In the end, however, terms like *black Anglos* and *white Anglos* seemed unwieldy, so I hope the reader will indulge me for the sake of readability. Of course, in using the term *Anglo,* I am not attempting to suggest all English-speaking whites were of British origin, only that English-speaking whites from the United States generally shared a common identity.

In speaking of the European immigrants, I generally use the word *German.* I am, of course, aware that the unified German state was not created until 1871 and that many who came from the region retained their regional identities, though my reading of the evidence as a whole points to the existence of a German identity throughout the period under examination. There was also a small Czech-speaking community on the Brazos. Where relevant, I distinguish between Czechs and Germans, but since the boundary between the two was often faint, I usually mean to include them when I speak of "the Germans."

This is a study of a five-county region (six since 1873, when the eastern half of Austin County was separated from the western half to form Waller County; see map 1: the counties shown are Brazoria, Fort Bend, Austin, Washington, and Grimes). Located to the west of Houston in what is now called East Cen-

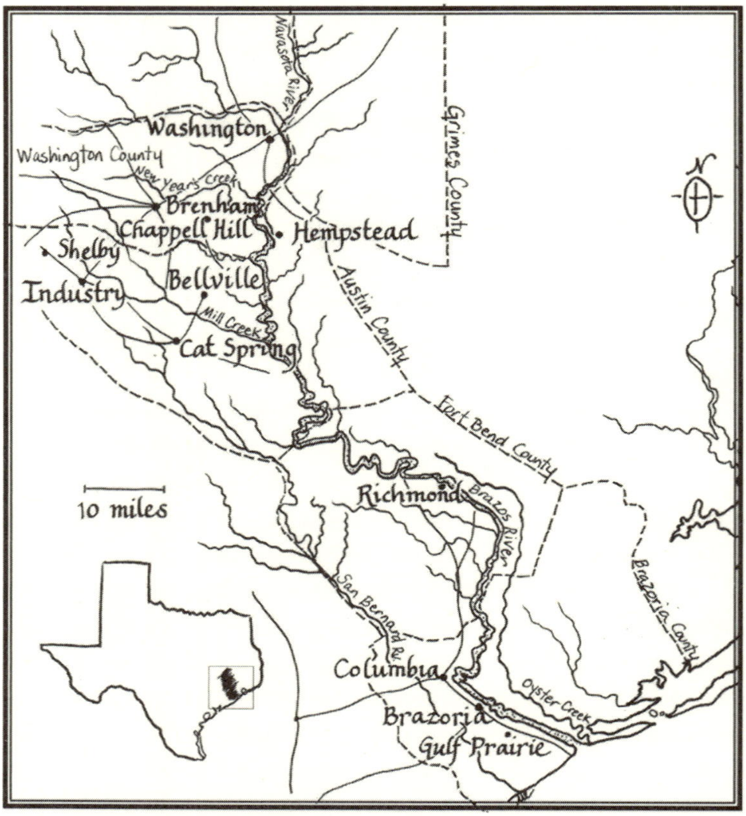

Map 1. The lower Brazos.
Map by Michelle Lerner.

tral Texas, these counties never actually comprised a political entity known as "the Brazos." There was, between 1834 and 1836, a "Department of the Brazos," but it encompassed territory not included in this study. The exclusion from the study of nearby counties on the Colorado River, many of which were economically and demographically similar to those of the Brazos, is admittedly somewhat arbitrary. Nevertheless, there was a measure of integrity to the five-county lower Brazos. Most obviously, they shared the river itself, and thus had an important artery of transportation and communication in common. Also, it turns out that early residents of the "Lower Country" (a contemporary name for present-day Fort Bend and Brazoria counties) had considerable speculations in "Upper Country" lands (Austin, Washington, and Grimes counties). Finally, the counties of the lower Brazos were usually drawn into legislative districts with each other, giving them a measure of political integrity.

Yet even a subregion like the Brazos encompassed a wide variety of localities, which contemporaries referred to as "neighborhoods." To ignore the region's mosaic of neighborhoods would be to miss a very important aspect of residents' lived experience. In order to capture this diversity, I have selected three neighborhoods for closer examination. The first is Gulf Prairie, located in the Lower Country on the Brazos River delta in present-day Brazoria County. Gulf Prairie was one of the oldest and wealthiest plantation areas in Texas, the heart of the state's "sugar bowl." The second neighborhood is Chappell Hill, located in the Upper Country in the Brazos river bottoms of present-day Washington County. Chappell Hill was also a wealthy plantation district, but it differed from Gulf Prairie in that it developed about fifteen years later and that its primary crop was cotton. The third locality is the Upper Country settlement of Cat Spring, situated in the hills of present-day Austin County, just outside the Brazos River bottom.[14] Cat Spring was one of the state's oldest predominantly German neighborhoods and continued so into the twentieth century. All three neighborhoods receive greater attention in chapter 1.

Our examination begins with a short prologue covering the years before 1821, when the Brazos was part of the Spanish borderlands. After 1821 the Brazos was transformed by four waves of migrants, the subject of chapter 1. Chapters 2, 3, and 4 treat the fundamental relationships of antebellum Brazos society, each emphasizing the difficulty with which they were transplanted to Texas, as well as the influence of the border. Husbands and wives, both free and enslaved, are the subject of chapter 2. Negotiations between masters and slaves are the subject of chapter 3, and connections between households, including connections in the realm of politics, are the subject of chapter 4. The presence of the Germans, whose true sentiments on the question of slavery were always in doubt, made neighborly and political relations on the Brazos extremely complex. The fifth chapter addresses the Civil War along the Brazos. With rampant slave resistance and actual violence between the German and Anglo communities, no conflict was more deserving of the name. Emancipation and the conclusion of the Civil War serve as a convenient bookend to this phase of the region's history. Although the tensions and conflicts of the antebellum period did not vanish, the Brazos became more "bordered land" than borderland. Since the beginning of Anglo settlement in the 1820s, the Brazos had been pulled by Mexico and the United States. With slavery gone, Mexico's attraction receded and the Brazos became more typically "southern." Aspects of the past survived, but a very different era dawned.

Prologue

BORDERLANDS

The river we now call the Brazos has had many names. Native Americans reportedly referred to it as Tokonohono. La Salle called it La Maligne, or the wicked one. Other names include La Trinidad, Santa Teresa y Barroso, Espíritu Santo, Río Rojo, Río de Señor San Pablo, Jesús Nazareno, San Gerónimo, and Baatse. Exactly how the Brazos River got its present name remains a mystery. A number of tales exist, some more plausible than others. The most popular one is that Coronado named it "Los Brazos de Dios" ("the arms of God") in 1541 when he encountered it near its source after crossing the arid Staked Plain of northwest Texas. Other tales involve sailors, lost at sea and in need of fresh water, tracing a muddy streak in the Gulf back to the river's mouth and naming it in gratitude for their deliverance. Whatever its name, the Brazos is the third-largest river in Texas, behind both the Rio Grande and the Red rivers, flowing like most of the state's major rivers in a northwest-southeast direction and emptying into the Gulf of Mexico. It winds its way slowly below the falls, which are located about three hundred miles from the mouth. Its muddy reddish-brown color is testament to its immense drainage area, some 41,700 square miles, which made its bottom and delta lands some of the most fertile in North America.[1]

To travelers approaching from the Gulf, the mouth of the Brazos was obscured by sandy barrier islands. Once past the beach, the land surrounding the mouth was table-flat for miles inland. Anglo settlers referred to this coastal plain as the Lower Country. At the start of the nineteenth century, the soil was covered with dense canebrakes, which towered over a man's head, even when he was mounted on horseback. Moving through the cane was an arduous process, made easier with a blade to hack away the stalks. With the reeds towering fifteen or more feet above the ground, a person walking though the cane traveled in darkness, as if in a tunnel. Outside the river bottoms, which varied from two to twenty miles wide, was an open coastal prairie, virtually treeless and covered with tall grass. In some parts the prairie extended for miles in all directions, broken up by clumps of trees that struck foreign observers as islands in a sea of grass. The horizon was circumscribed by the forests that marched along the creeks and rivers. Travel was easier on the prairie, but people could also easily become disoriented without landmarks.[2]

About sixty miles from the coast, the flat surface gave way to the gently rolling hills of what Anglo settlers later called the Upper Country. Here the canebrakes were more sparse and were supplanted in the bottoms by a semi-tropical, moss-draped hardwood forest that provided a haven for a variety of wildlife, including alligators, poisonous snakes, and wildcats. Transit through the bottoms was just as difficult as through the cane due to the density and darkness, and for seven months of the year, the air was heavy, humid, and swarming with flies and mosquitoes. The various tributaries of the Brazos—the Navasota River and New Year's, Mill, and Oyster creeks—supported a similar ecology, on a smaller scale. Between these streams, in the uplands outside the river bottoms, post-oak savannah dominated. The occasional herd of bison found its way onto these prairies, and sometime after the arrival of the Spanish, a large herd of wild mustangs took up residence.

The Karankawa people had mastered the art of living in the Lower Country. Food was plentiful, but getting enough required expert knowledge of the seasonal patterns of fish and game as well as fruits, berries, and roots. The Karankawas had worked the coast and rivers much the same way for generations. During winter and spring, when redfish and black drum abounded in the estuaries and lagoons between the barrier islands and the coast, the Karankawas gathered in settlements numbering up to four hundred to fish. But during the summer, when the fish population decreased, the Karankawas broke into much smaller groups, dispersed, and moved up to twenty-five miles inland to hunt white-tailed deer and bison on the prairies and gather hackberries and mustang grapes from river bottoms. As the Karankawas migrated seasonally to and from the coast, Tonkawa Indians shadowed their migration, taking up residence twenty-five miles from the coast during the winter when the Karankawas were on the coast and moving farther inland for the summer. The two groups, who seem from archaeological evidence to be culturally distinct, may have engaged in communal hunting expeditions and other kinds of exchange. The Brazos was a borderland well before the Europeans arrived.[3]

The invasion of Texas by Spain created a new point of contact for the Karankawas. From the sixteenth to the early nineteenth century, Texas was on the far periphery of the colony of New Spain. Spanish colonial officials had little interest in the economic potential of the area, and the ties to the larger economic system, while present, were irregular and thin. In fact, Spain's chief interest in Texas had little to do with trade—Texas was seen primarily as a military buffer to keep the French and later the English, and especially the *indios bárbaros*, from threatening more valuable parts of the empire. During the late seventeenth century, as French traders and Jesuit missionaries traveled through the Mississippi Valley, Spanish officials began to worry about the security of their northern province. Reports that priests and traders were forming alliances with Caddoan and Atakapan groups around the Sabine River were alarming enough, but news of a French fort near Matagorda Bay, just down the coast from the mouth of the Brazos, impressed Spanish officials as a serious challenge. They immediately dispatched a force to dismantle the fort and capture the interlopers, only to find that the Karankawas had beaten them to the punch and killed the soldiers themselves.[4]

With the destruction of the French garrison, the Karankawas served notice that they would contest European efforts to make Texas part of the Spanish or French empire. The Spanish method of extending imperial dominion was to incorporate native groups into the empire under the temporal authority of the Crown and the spiritual authority of the Roman Catholic Church. The Spanish tried, with mixed success, to herd Karankawas into missions during the eighteenth century. Relations between the Spanish missionaries and the Karankawas soon soured, and the two groups embarked on a low-level war with each other. Despite Spanish resolve to "exterminate" the savages, the Karankawas proved a difficult adversary.[5] Their knowledge of the barrier islands, bays, marshes, and canebrakes along the coast served them well, and Spanish forces found themselves routinely frustrated in their attempts to punish Indian "depredations."[6]

In fact, Karankawa "depredations" against Spanish cattle herds were nothing of the kind; rather, they were a strategic adaptation to changes in the Karankawas' ecological world. One of the missions' principal economic activities was herding cattle and then driving them to market in French (later Spanish) Louisiana. Over time, the cattle herds grew, dispersed, and crowded out the herds of bison upon which the Karankawas depended during the summer.

To make up for the loss of the bison, the Karankawas simply substituted mission cattle, a large portion of which were not even branded. Killing the cattle brought retaliation by Spanish authorities, and the conflicts easily escalated, but for the Karankawas the choice was a simple one: they could maintain their culture, traditions, and seasonal way of life by eating beef instead of bison, or they could abandon the strategies that had served them so well for generations.[7]

In spite of the periodic conflict over livestock, the relationship between the Karankawas and the Spanish was not one of unremitting hostility. Since the 1680s the Karankawas' world had been in constant motion, and they were remarkably light on their feet, devising numerous strategies for coping with new problems as they arose. In the middle of the eighteenth century, a new problem appeared to compound the thinning of the bison herds. The Comanches, a large and militarily powerful group from the Great Plains, had begun pressing southward at the expense of inland groups and may even have begun periodic hostilities with the Karankawas. Recognizing the military threat and the ecological problem of poorer hunting, the Karankawas turned to the Franciscan friars. At the missions they could find a haven from Comanche incursions as well as food during the summer. The friars' price was steep: Indians were to take religious instruction, abandon indigenous beliefs, and learn Spanish farming techniques. Those who tried to leave could be "corrected" by whipping or stocks.[8]

The Karankawas' primary goal was to maintain their cultural integrity while using the Franciscans to solve the problems of invasion and ecological change. And by holding out the prospect of coming to the missions, the Karankawas were largely able to negotiate the terms of cooperation. For example, the friars found the Indians reluctant to come to Mission Espiritu Santo and Mission Rosario since they were too far inland. The Karankawas expressed the wish that one be built closer to the coast, which seems to be the reason behind the founding of Mission Refugio in 1793.[9] Far from becoming dependent on the Spanish, the Karankawas seem to have come and gone seasonally, with the number of visits peaking during the summer, indicating that the missions had merely become integrated into the Karankawas' pattern of coastal-inland seasonal migration. The Spanish might have considered their efforts a success if the Indians had wholly accepted Christianity, but here again the Karankawas appropriated Spanish offerings with the goal of strengthening their own independence. The Franciscans had baptized two hundred Karankawas by 1768, but few stayed long in the missions, and apostates frequently left with the added skill of fluency in Spanish.[10]

The Spanish brought other, deadlier problems to the Karankawas in the form of disease. For Indian communities throughout the hemisphere, contact with Europeans and Africans exposed large numbers of people to pathogens for which they lacked immunity. For the Karankawas, the process seems to have begun almost immediately, with the landing of Alvar Núñez Cabeza de Vaca in 1528. After a great number of Indians died of a stomach ailment, they blamed the Spanish castaways and seemed ready to put them to death in retaliation, but the Karankawas backed down when some of the Europeans became sick as well. La Salle's men reintroduced European pathogens in the seventeenth century, but disease took a far more serious toll in the eighteenth century. From 1749 to 1778 there were at least three epidemics of smallpox and measles that carried off a large percentage of the Karankawa population. From a total of about 8,000 men, women, and children in the 1680s, the Karankawa population dropped to about 3,000 in 1780. By 1820, the eve of Anglo settlement, there were probably a little more than 2,000 Karankawas living on the Gulf coast between present-day Galveston and Corpus Christi. Of these, about 400 were Cocos, the subgroup that lived near the Brazos.[11]

As devastating as disease was, the Karankawas adapted to the population decrease as readily as they had to the drop in the number of bison. Again, hoping to avoid compromising their lifeways by moving or becoming dependent on the missions, they devised a strategy to make the most of a bad situation. In this case, they reorganized the subtribal groups that had structured Karankawa life for decades. Before the late eighteenth century, there had been five major groups spread out along the coast. After the 1780s some of the more depleted groups appear to have merged and intermarried with others. In addition, the Karankawas seem to have incorporated some Tonkawas, who generally lived a hundred or more miles upriver, close to the falls of the Brazos.[12] The new alignment allowed the Karankawas to continue the economy of seasonal migration and gave them the political and military strength to resist Spanish and other Indian aggression. In short, the Karankawas had mastered the art of maintaining their cultural identity through creative adaptation.[13]

Events of the new century challenged both the Karankawas' and the Tonkawas' adaptive skills. The region that had for two centuries been a Spanish imperial backwater now attracted the attention of Anglo speculators and planters. In 1806, Anglo American trading along the Red River in East Texas nearly brought about armed conflict between Spain and the United States, which claimed Texas as part of the Louisiana Purchase. Open hostilities were averted only by the conclusion of the Neutral Ground Agreement, under which the United States would hold its military forces east of Arroyo Hondo, just west of Natchitoches, Louisiana, and Spain would keep its troops west of the Sabine River.[14]

After 1807, when Zebulon Pike's expedition was expelled from Spanish territory and brought back descriptions of fertile lands and a mild climate, various groups in the United States began to ponder ways of getting access to Texas. Political turmoil in connection with Mexico's independence movement supplied an opportunity. The first to take advantage of the situation was José Bernardo Gutiérrez de Lara. A revolutionary whose target was the royalist administration in San Antonio, Gutiérrez assembled an invasion force in Louisiana in 1811–12. His force included an Anglo contingent, whose members, led by Augustus William Magee, were motivated less by a desire to secure a republican Mexico than by visions of landownership in Texas. The expedition succeeded in overthrowing the royalist governor at San Antonio, though Magee died along the way. Gutiérrez, whose caudillo tendencies alarmed many of his followers, was soon overthrown by a U.S.-backed general, who was in turn defeated by royalist troops later in 1813.[15] In 1819–21, James Long, another Mississippian, mounted two failed expeditions to Texas. Though he did not actually make it to the Brazos, his expeditions demonstrate Anglo American interest in Texas. For the Brazos region, the political events of the independence movement were far less significant than the fact that a large number of Anglo Americans had gained knowledge of Spanish territory.

In 1821, with Anglo American designs on Texas now painfully clear, Spain adopted a new policy. Rather than attempt to hold the line against migrants from the United States, it now encouraged them to settle through grants of land and in the process hoped to strengthen its hold on Texas. Among the first to respond to the new policy was Moses Austin, a Connecticut native who had moved to Spanish Louisiana (present-day Missouri) in the 1790s. A man of the borderlands himself, Austin understood the attraction Texas held for many in the United States, and he knew how to turn it to his own advantage. In 1820 Austin traveled to San Antonio, where he lobbied for the passage of a law allowing for the colonization of three hundred Roman Catholic families along the Brazos and Colorado Rivers. He ultimately succeeded but died soon after returning from Texas. His son Stephen, hoping to unencumber the family estate, decided to run his father's venture himself, but Mexico's newly independent provisional government revoked Moses Austin's contract, preferring to regulate colonization under a general law.[16]

In early 1823 such a law was passed by the government of Emperor Agustín de Iturbide, and Stephen F. Austin's contract was confirmed. The law contained most of the same provisions that his father's contract had. It allowed Stephen F. Austin, as *empresario,* to settle three hundred families on land surrounding the Brazos and Colorado rivers. Each settler was entitled to land, consisting of a league (4,428 acres, also known as a *sitio*) for those who intended to raise stock and a *labor* (177 acres) for those who intended to farm. Other provisions required that settlers develop the land within two years, addressed the founding of towns, regulated the empresario's "premium lands," set the colonists' tithes and taxes, and provided for their naturalization. The law also banned the sale or purchase of slaves and provided for the emancipation at age fourteen of all slaves born within the empire. The Imperial Colonization Law of 1823 did not, however, make slavery itself explicitly illegal. Settlers were additionally expected to profess Roman Catholicism, abide by the laws of Mexico, and pay nominal processing fees. Shortly after Austin received the contract, however, Iturbide was deposed and all legislation passed under his auspices nullified. After more lobbying, Austin managed to confirm his grant one more time, and he departed from Mexico City for Texas. In the succeeding years, both the state and federal governments modified the colonization policy, but the outline remained substantially the same. For residents of the United States, who could no longer purchase federal lands in parcels smaller than 640 acres, Mexico's offer was attractive. The migrants soon came.[17]

Chapter 1

MIGRATIONS

Humorist Joseph Glover Baldwin has left us the most enduring sketch of the Brazos migrant. In an 1853 piece titled "How the Times Served the Virginians," he contrasted two types of Mississippi planters during the "Flush Times" of the 1830s. The first was the well-heeled and honorable Virginia gentleman who moved to the Magnolia State and haplessly lost the family fortune to gamblers, speculators, and other crooks. The second was the wily southwesterner, a rather coarse man on the make, scarcely different from the predators who preyed upon the Virginians. In Baldwin's telling, the latter was the archetype of the Brazos planter. Hounded perpetually by his creditors, he sought relief through such ruses as feigned bankruptcy and fraudulent property conveyances. When those tricks failed, he "[sent] out coffins to the graveyard, with negroes inside, carried off by sudden spells of disease, to be 'resurrected' in due time, grinning, along the banks of the Brazos."[1] Baldwin's sketch was, of course, overdrawn and played to the racial stereotypes of the era, but its core contained a kernel of truth. Plantation agriculture made and unmade a great many fortunes. For the thousands of planters and farmers who experienced financial death in their home states (and for those who had never been financially "born"), a move to the Brazos promised financial resurrection. For the thousands of African Americans who accompanied them, it promised not resurrection, as in Baldwin's telling, but the continued "social death" of slavery.

For all its color, Baldwin's sketch missed the complexity of the Brazos migrations. In fact, no fewer than four major migrations (more, if we choose to dissect them further) made the Brazos one of the most demographically complex areas of the American South. The first and most statistically significant was the movement of Anglos and enslaved African Americans from the southern United States beginning in the 1820s and increasing steadily through the antebellum period. As we will see, this migration was actually three, comprising the initial migrants, who built their fortunes on Mexican land grants, mostly in the Lower Country; the later migrants, who purchased their lands from the grantees and settled mostly in the Upper Country; and the slaves that each group brought with them. The second major migration involved more than six thousand African Americans who were sold to the Brazos on the interregional slave market from the slave-exporting states of the upper and the seaboard South. Most of these came initially to New Orleans, but some came through the lesser slave markets of Houston and Galveston. Because these migrants arrived as property of slave-trading firms and not as part of a master's retinue, they deserve consideration apart from the planter migrations. The third major group of travelers consisted of about three thousand German-speakers from central Europe. The earliest German settlements appeared on the Brazos in the 1830s, and the numbers increased slowly but steadily before accelerating rapidly during the 1850s. The fourth group of migrants endured the longest and most painful journey to the Brazos. They were Africans, born free but enslaved and shipped to Cuba during adolescence or early adulthood. In the years surrounding the Texas Revolution, Brazos planters financed the transportation of several hundred Africans from Cuba to Texas, where their influence belied their relatively small numbers.

These migrations shared much in common. All were in some way connected to the expansion of the Atlantic economy. What historian Eric Hobsbawm has termed "the dual revolutions" transformed Europe and the New World in complex ways. The rise of industry in Europe and the northeastern United States stoked a demand for New World products. Some of these, such as cotton and indigo, supplied the textile mills that drove the early industrial revolution. Other products, such as sugar, tobacco, rice, and coffee, fed or stimulated a growing population of peasants-cum-wage-laborers, no longer able to produce their own food. The demand for these products led to the geographic expansion of slavery into the continental interior of both North and South America, often fueled by an illegal African slave trade. All of these migrations were also in some way affected by the border politics of Mexico and the United States, from the Mexican government's initial invitation to Anglo colonists to the skillful exploitation of border conflict to import African slaves. Even the presence of the Germans was partly due to the fact that Mexican authorities saw them as an attractive alternative to the troublesome Anglos. The result was a babel of languages, labor regimes, folkways, and cosmologies—in short, a typically heterogeneous borderland.

Resurrection along the Brazos

Anglo and African American migration to the Brazos drew, predictably, on the U.S. South. It was there that the cotton boom had fueled countless dreams of wealth and independence, and there that so many of those dreams died in the wake of the financial panics of 1819 and 1837. For thousands of would-be landowners, Mexico's policy of granting colonists a league and a *sitio* totaling more than four thousand acres offered a way to acquire—or reacquire—something that was increasingly unattainable in the United States. In the 1820s and 1830s, a sizable portion of those who settled in Austin's colony came from the states nearest Texas. Of the heads of household in 1826, 29 percent came from Louisiana, with Missouri, Arkansas, Mississippi, Tennessee, and Alabama contributing percentages in the single digits. However, by 1835 the sources of migration had changed. At 14 percent, Louisianans no longer dominated, while the percentages of Alabamians and Tennesseans rose to 12 and 14 percent, respectively. For the state as a whole, this later pattern grew more pronounced. By 1860, Alabama accounted for 15 percent of all non-Texas-born natives and Tennessee for 18 percent, with Georgia contributing 10 percent and Mississippi 9 percent. The remainder was composed mostly of southerners, with a few northern-born settlers scattered throughout.[2]

In most cases, migrants came directly from their states of origin to Texas. In his detailed study of East Texas (that is, east of the Trinity River, a region that does not include the Brazos), historian Barnes Lathrop found that at least one-fourth, and possibly one-third, of all migrants made an intermediate stop on the way to Texas. A close examination of one community in Washington County suggests a similar pattern for the Brazos, though to a slightly lesser degree. Census returns detailing children's birthplaces show that about 22 percent of the 1860 households in Chappell Hill had lived in some other state before coming to Texas. Because the method does not catch families who did not have children in the intermediate state, the actual percentage was certainly higher. Not surprisingly, those from farthest away were more likely to have spent time someplace in between. Nearly three-fourths of the migrants from Georgia, South Carolina, and Virginia made intermediate stops, which were split about evenly between Alabama and Mississippi. Kentuckians, Alabamians, and Tennesseans tended to come directly.[3]

Antebellum migrants followed two principal routes to the Brazos, one overland and one by sea. Most probably came overland, usually via Natchitoches, Louisiana, through the Mexican outpost of Nacogdoches, then down the muddy San Antonio Road toward the Brazos. They met the river at the army post at Tenoxtitlán (founded 1830 in order to establish a greater Mexican state presence in Central Texas and falling to ruin after 1836), then trundled downstream toward the Austin Colony. Alternatively, migrants might head southwest at the Trinity River in East Texas, which took them to the Brazos in present-day Washington County. The latter was the route taken by the family of John Davidson from North Carolina in 1843. Their farm threatened during the depression of the 1840s, the family decided to relocate. Taking full advantage of kin and friends who had left years earlier, the family spent more than a year migrating to Texas, with stops as far afield as Illinois. They eventually reached Washington County in 1844, though illness forced them to leave after six months.[4]

Still other migrants sailed to the Brazos through New Orleans. Often these included large-scale slaveholders, who preferred to pay for the quick passage of their families and laborers to the Brazos rather than expose them to the risks of the overland journey. Hamblin Bass, who purchased Morgan Smith's massive Brazoria County sugar plantation in 1860, traveled with his entire contingent of more than one hundred laborers from Barbour County, Alabama, to Mobile and then sailed from there to New Orleans on one of the coldest days of the year. They reached New Orleans the next morning with the ship's deck encased in an inch-thick sheet of ice. After enduring a long, cold night in the basement of a nearby house, the party crossed over to Algiers on the west bank of the Mississippi, took a train to Burwick Bay, then embarked on a steamer for Galveston. Among the passengers were more than 238 slaves, divided among twelve white families. Thomas Blackshear, a Georgia planter who settled in Grimes County in 1858, steamed from Tallahassee to New Orleans, from New Orleans to Galveston, and from Galveston to Houston with only a handful of his slaves. The party then proceeded on foot to their final destination near Navasota. The majority of Blackshear's slaves appear to have traveled overland the entire way and did not arrive until months afterward.

Railroads played only a minor role in channeling antebellum migration to the Brazos. Not built until the later 1850s, they primarily shipped crops to Galveston and Houston and did not connect with the national network until after the Civil War.[5]

Historians have frequently treated the migration of slaves into Texas as part of a larger process of southern cultural diffusion. Anglo southerners, runs the argument, brought certain practices—agricultural methods, architecture, and food ways—with them as they migrated westward. Slavery, then, stands beside corn culture, dogtrot houses, and hoe cake as yet another practice unthinkingly transplanted to Texas. But while slaveholding certainly worked its way into Anglo-southern culture, it does not follow that slaveholding culture caused large-scale slavery. Most Anglo southerners, in Texas and at large, were nonslaveholders. Conversely, natives of Europe and of the northern United States were among the largest slaveholding planters along the Brazos. Slavery and nineteenth-century Anglo-southern culture may have been closely related, but they were not identical.

In fact, slavery had much less to do with folk custom than it did with a demand for labor. Most historians would confirm the correlation between slavery, readily available land, and the desire for labor, usually deployed toward producing cash crops for sale on foreign markets. David Eltis's summation of the situation for the seventeenth- and eighteenth-century Caribbean applies just as well to nineteenth-century Texas: "The master-slave relationship became the norm because of the scarcity of labour. Without coercion there would have been little production for export in a land-abundant environment." We might refine the statement to say that plantation slavery tended to take root in areas with good soils (or other natural resources, such as minerals) and reasonably good transportation connections and where other forms of labor, such as indentured or wage labor, were not practical. Had there been no market for cotton and sugar, or had a large population of potential free laborers existed in Texas, there would have been little demand for slaves.[6]

The difference between saying that Anglo southerners brought slavery to Texas and that slavery spread into Texas as part of a general nineteenth-century expansion of plantation agriculture may seem inconsequential—after all, slaves wound up in Texas either way—but it has important implications for how we understand slavery and the history of the region. To subsume the migration of slaves under the heading of Anglo migration is to endorse, tacitly, the notion of an organic master-slave relationship, when in fact the relationship was anything but that. Many enslaved migrants fled their masters, either en route or shortly after arrival. In addition, because slaves were legally defined as chattels and economically defined as capital, their migration into Texas was considerably more multifaceted than Anglo migration. Finally, conceptually fusing African American with Anglo migration obscures the fact that similar migrations were occurring throughout plantation America, all caused by increasing consumption of plantation products. In Brazil, slaves moved from the older sugar-growing regions of the Northeast into the coffee-growing areas of the Southwest. In Cuba, older coffee and tobacco farms, run with little or no slave labor, were transformed into large sugar *ingenios*, with the corresponding movement of slaves. From this perspective, slave migration and plantation agriculture appear less unique, less particular to any specific culture, and more connected to long-term shifts in the Atlantic economy.[7]

Anglo migration to the Brazos drew on the South at large, but not randomly. The local neighborhoods that formed in Texas differed significantly from one another, shaped by quirks of local geography and politics and by interregional networks of information and settlers. For a finer-grained portrait, we will examine two Brazos neighborhoods, both destinations for southern migrants, both significant planting regions. These commonalities aside, they were quite different, with one, Gulf Prairie, becoming a center of sugar production, and the other, Chappell Hill, becoming a major producer of cotton.

It is nearly impossible to locate Gulf Prairie on a map today. No town of that name exists or, for that matter, ever existed. Equipped with a very detailed map, the sharp-eyed reader might discern a small cross-shaped symbol representing Gulf Prairie Presbyterian Church on the west bank of the Brazos off State Highway 36 near present-day Jones Creek, only fourteen miles from the mouth of the river.[8] The church sits on what was once Peach Point Plantation, one of seven large estates that formed what locals all knew as the Gulf Prairie neighborhood. As Texas's first true plantation neighborhood, Gulf Prairie was atypical in many respects. Unlike most other Texas planters, who purchased their plantations from speculators and earlier arrivals, the Gulf Prairie planters generally received their land in grants from the Mexican government. The Gulf Prairie planters were also a much more stable and enduring cohort than others. All of the seven principal plantations in 1832 were still in the hands of the original families at the close of the Civil War. Gulf Prairie's plantations were also unusual for their size, crop mix, and slave populations. The seven main plantations averaged more than seventy slave laborers, who, beginning in the 1840s, actually grew more sugarcane than cotton. Finally, the laborers in Gulf Prairie included a large number of illegally imported African-born slaves; for a few years, these made up almost half of the local slave population and constituted what was likely the largest rural community of Africans in the antebellum South.

Most of the Gulf Prairie planters received their lands courtesy of a Mexican state that saw settlers as a bulwark against encroachment by Anglo filibusters and *indios bárbaros*. In a near-perfect example of what historian Patricia Limerick has termed "first in time, first in right," fifty-six individuals filed claims to the land as soon as they could be registered in 1824. The land they claimed was the most valuable in Texas, not only because of its obvious fertility but because of its location near the mouths of the Brazos and San Bernard rivers, which made transportation reliable. Although both rivers were navigable well above Gulf Prairie in all seasons, the Brazos followed a twisted, tortured course on its way to the Gulf. A farm near the mouth might actually be days closer to New Orleans than one only a few miles due north. Well aware of these factors and hoping to profit as much as possible from their lands, claimants snapped up Lower Country riverfront tracts immediately. In present-day Brazoria County, Anglo settlers patented fifty-six parcels of land between July and August 1824, leaving not a single tract fronting the Brazos, the San Bernard, or Oyster Creek. River-bottom leagues more than a hundred miles above the point of year-round navigation vanished just as quickly. The major exception was an enormous seven-and-a half-league (33,330-acre) tract on Gulf Prairie, which the Mexican government allowed Austin as his premium for undertaking colonization. In the succeeding years, Austin and his heirs alienated pieces of the tract, furnishing land for the richest plantations in Texas.[9]

The contemplated plantations of Gulf Prairie sat on land belonging—by right of usage—to the Cocos band of Karankawa Indians. To colonists who coveted their lands, as well as to Mexican authorities, the Karankawas were savages whose presence not only impeded the future civilization of the wilderness but constituted a threat to the borders of established civilized enterprises. The Karankawas occupied land that was clearly destined to be redeemed through agriculture, and in addition their very savagery—evident in their dress, methods of sustenance, and lack of proper gender hierarchy—constituted an affront to civilization itself. After a series of confrontations, Austin determined in 1824 to wage a war of expulsion. With the blessings of Mexican authorities in Saltillo, he personally led an expedition against the indigenous residents of the lower Brazos and Colorado rivers. Participants included would-be planters, adventurers, and thirty enslaved residents of Jared Groce's Bernardo Plantation from the Upper Country. The force managed to chase the Karankawas westward to La Bahía, where Mexican military and religious leaders brokered an agreement by which the Cocos pledged to stay well west of Austin's colony. The agreement, however, forced the Karankawas into territory controlled by Lipans and Comanches, whose raids had long plagued them. In 1825 and 1826, small groups of Karankawas attempted to return to their familiar territory on the Brazos and the Colorado, only to become subject to a shoot-on-sight order from Austin. Later, with the killing of two Anglo settlers in 1826, colonists trapped a group of Karankawa men, women, and children near the mouth of the Colorado River and massacred a score or more. From this point forward, Gulf Prairie settlers had no trouble from the indigenous former residents of their lands.[10]

Who were the Gulf Prairie planters? To generalize, they were early comers from the Upper South with a canny understanding of how to use international borders to their best advantage. At least one had prior borderlands experience, and several had mercantile experience, which they used to transform their land grants into large slave labor forces. Their numbers are small enough that we can meet each of them. The first Anglo planter on Gulf Prairie was a Kentuckian named John McNeel, who arrived in 1822 and patented a league that fronted the San Bernard River, stretching back onto Gulf Prairie (see map 2). He brought between twenty and twenty-five slaves to Texas, indispensable in this region of abundant land. McNeel apparently ran into difficulty early on, as evinced by his pleas to Austin to let him delay payment of his recording fees for having pledged the year's crop to a commission merchant, but his fortunes turned around in the next few years. In 1831 a traveler found him prosperous; McNeel had recently sold a five-thousand-dollar cotton crop. To the visitor, McNeel appeared the wilderness patriarch, living in Edenic simplicity and abundance in "a good log house, just on the verge of a fine grove, partly shaded by China trees, newly planted before it for ornament, and overlooking his whole domain."[11] Three of McNeel's adult sons, John Greenville, Pleasant, and Leander, also claimed tracts on the San Bernard side of Gulf Prairie, which

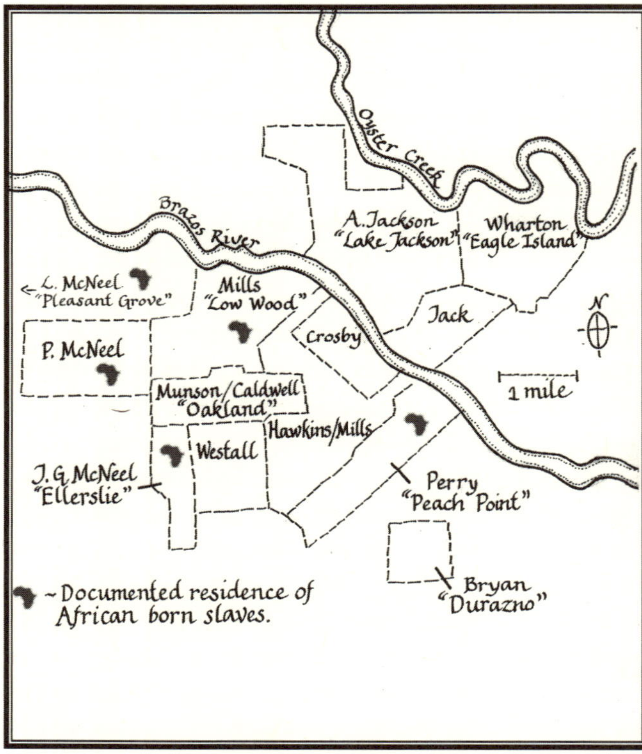

Map 2. Gulf Prairie plantations.
Map by Michelle Lerner.

they eventually developed into sizable plantations. A fourth son, Sterling, became a planter as well, establishing himself a few miles up the Brazos.[12]

For those without a large slave labor force, the Mexican borderlands provided other opportunities for entry into the planter class, as the experience of McNeel's neighbor, Thomas Westall, illustrates. Unlike McNeel, Westall was not a particularly wealthy man, arriving in Texas from Tennessee in 1824 with four slaves. Shortly afterward, he patented more than nine thousand acres in present-day Wharton, Austin, and Fort Bend counties, as well as several hundred acres on Gulf Prairie, which he chose as his future plantation. Westall was short on cash and had difficulty paying the required fees, a problem he solved by hiring out his slaves to Stephen F. Austin at $180 for the year (although shortly afterward he took one back). In the fall of 1824, Westall returned to Tennessee to get his family. While there he formed a small commercial concern with the help of an associate's brother and purchased between two thousand and three thousand dollars' worth of merchandise. He may also have used his lands as collateral. In any event, Westall apparently had no desire to run a store on a permanent basis; rather, he planned to sell his stock as quickly as possible in "Spanish Towns" and use the proceeds to purchase mules. The record does not reveal whether he actually traded in any of the Hispanic areas of Texas, but he did return and sell a wide variety of merchandise, including cloth, tobacco, and wine, to Anglo residents.[13]

Westall used his mercantile profits and land grants to purchase more land and slaves. When he died in 1833, only nine years after coming to Texas, he left an estate valued at more than twenty-eight thousand dollars. His real estate holdings had increased to include, among other acquisitions, another tract on Gulf Prairie and several town lots. His five slaves had become eighteen. His home furnishings were comfortable: he had glass windows, a mirror, silverware, and a set of encyclopedias. Agricultural implements, however, made up the bulk of his nonhuman personal property, suggestive of the true nature of his move to Texas. Indeed, by 1850 under his son Andrew, the plantation boasted a labor force of forty-seven and had converted from cotton to sugarcane production.[14]

Another Gulf Prairie family, the Munsons of Mississippi, had experience in Texas that predated Austin. The patriarch, Henry William Munson, initially sought his fortune in the borderlands as part of the Gutiérrez-Magee expedition and passed through the region in 1813. Munson returned in the 1820s, initially settling in East Texas and then, dissatisfied with his situation there, purchased part of Austin's premium land in 1828. With seventeen slaves, Munson managed to accumulate several thousand dollars. After his death in 1833, Munson's widow Anne married James P. Caldwell, who continued running the plantation with a labor force of about thirty-five slaves. In the early 1850s William B. Munson assumed control of what was now known as Oakland Plantation, and in 1860, the plantation went to the heirs, with Munson's son Girard operating the plantation in the interest of the others.

For those who came too late to receive grants on Gulf Prairie, financial acumen and family connections could compensate. The brothers Robert and David G. Mills of Kentucky are a case in point. Described by an agent of the

R. G. Dun credit-rating company as "businessmen in every respect," the two came to Texas from Kentucky in the early 1830s. At first, the Mills brothers acted as commission merchants, dispensing such items as shoes, tobacco, cloth, and laudanum to Brazos farmers and planters and selling their customers' cotton on commission. Having arrived after the first Brazos land rush, Robert and David Mills did not receive the same massive land grants as their predecessors, although they did receive more than one thousand acres well west of the prime plantation land on the Brazos from the Mexican government. Instead, through their mercantile efforts they acquired land that had been granted to others. Contrary to a local legend that the brothers never foreclosed, a glance at the county court docket reveals no hesitation on their part in initiating legal proceedings; sometimes they accepted land or slaves as payment. Their plantation holdings grew rapidly and included Low Wood, a Gulf Prairie sugar plantation with nearly two hundred slaves, the largest in Texas. On the eve of the Civil War, Robert and David Mills owned three plantations in Brazoria County and a fourth in neighboring Matagorda, and they paid taxes on 430 slaves. But the Mills brothers' activities extended beyond the plantation trade. They also shipped regularly to Laredo, where their wares made their way deeper into Mexico.[15]

Rounding out the Gulf Prairie elite were members of the Perry-Bryan family. Like the Mills brothers, the Perrys and the Bryans made up for their relative tardiness by being well connected—in this case, to Stephen F. Austin. Emily Austin Bryan Perry was the empresario's sister, as well as the wife of James F. Perry and the mother, in a previous marriage, of Guy M., William "Joel," and Moses Austin Bryan. The Perrys and the Bryans came to Texas with eight slaves in 1831 at the urging of Stephen F. Austin. Although they arrived in time to receive valuable land grants in the Upper Country, the most desirable lands in Gulf Prairie had been spoken for. Austin remedied the situation by deeding almost nineteen hundred acres, known as the Peach Point tract, to James and Emily Perry. After a disappointing stint as a merchant in the administrative town of San Felipe and as the operator of a stock farm, Perry began planting full-time at Peach Point in 1832. Austin dispensed lands to the Bryans as well, with Joel receiving a tract on Gulf Prairie that he later developed into Durazno Plantation.[16]

The vast majority of Gulf Prairie's population resided on one of these seven plantations: the McNeels' Ellerslie, Pleasant Grove, and Pleasant McNeel plantations; the Westall Plantation; the Munson/Caldwells' Oakland Plantation; the Mills brothers' Low Wood; and the Perrys' Peach Point. The lineup changed little for most of the century, the only enduring alterations coming through succession upon the death of an owner and the addition of William "Joel" Bryan's Durazno Plantation in 1840. Durazno, however, took time to develop as a plantation, concentrating on livestock for the first several years of its existence. Other longtime members of the community included the Crosby family, who owned what is sometimes referred to as a plantation adjacent to Peach Point, but they focused more on livestock and river commerce than on raising cotton or cane. A small number of additional planters appear in the records for short periods of time, but the above-mentioned plantations formed the stable core of the Gulf Prairie community. There were also a few small-scale farmers, although most of the other free residents lived and worked on the plantations as overseers, artisans, bookkeepers, and sugar-makers. In 1860 there were only thirty-two free households in the entire neighborhood, and enslaved residents outnumbered free residents 802 to 320.[17]

For the Gulf Prairie planters, the link between Mexican land grants and slave acquisition was strong. Unlike most other Brazos planters, the Gulf Prairie group began with a great deal of land and parlayed it into ever-larger slaveholdings. In 1837, the Gulf Prairie planter families owned more than 150,000 acres of land throughout Texas. The Perrys and the Bryans seem to have been especially adept at leveraging their landholdings (usually their Upper Country lands rather than their Gulf Prairie lands) into slaveholdings. They might do so in various ways: by using land as collateral for loans, which they then used to purchase slaves; by selling land to finance the purchase of laborers and plantation supplies; or by trading land straight up for slaves. Even if a planter did not actively use landholdings to buy slaves, credit-rating agencies like the R. Dun Company took their lands into account in evaluating their credit-worthiness. In sum, the Gulf Prairie planters rather methodically transformed land grants into cash and credit, which they then used to build their plantations. They had sought and found their fortunes in the borderlands.[18]

The Upper Country followed a different migration pattern, one that more closely resembled the other Anglo-dominated regions of Texas. Although the same land policies applied in the eventual counties of Austin, Washington, and Grimes, the grants did not spur the immediate rise of a planter class as they did in the Lower Country. Very few of the early Upper Country grantees held on to their lands for an extended period of time. Because there was less reliable water access and the bottomland, though fertile, was not as easily cleared as in the Brazos delta, prices remained low and populations more sparse. These circumstances combined, as Nestor Clay told his father back in Kentucky, to make the Upper Country a "free fighting, stock raising, money hunting country," with the latter term encompassing small farming and land speculation.[19]

The small-farming and stock-raising economy did not last. Following annexation in 1845 and then picking up rapidly in the 1850s, plantation agriculture overtook the Upper Country. The size distribution of Washington County landholdings illustrates the process, as large grants suitable for ranching were subdivided for cotton planting. In 1837, 77 percent of all landowners owned more than one thousand acres each; in 1860, only 14 percent did (see table 2). Over the same period, the percentage of slaveholding households jumped from less than 10 percent to 52 percent. Even free-fighting, stock-raising Nestor Clay was changing his money-hunting methods. The owner of a slave named Nathan, Clay initially deemed it more profitable to hire him out to a Gulf Prairie cotton planter than to employ him as a herdsman. In 1833, Clay changed his mind and asked for Nathan's return, as he was "about putting in a crop of cotton & cant spare him longer."[20] Many of the original Upper Country landowners, with fewer slaves at the outset and with poorer river transport than their Lower Country counterparts, could not afford to wait until their lands grew valuable enough to buy laborers, so they simply sold out. In 1860 one longtime resident lamented, "Our country is rapidly filling up, with wealthy farmers. A great portion of the prairie—and it had no fields in it worth naming . . . are now under cultivation."[21]

A closer look at one Upper Country neighborhood, Chappell Hill in Washington County, affords us a fine-grained portrait of this transformation. Situated at the point where New Year's Creek flowed into the Brazos, 393 miles from the mouth by river but only 90 miles as the crow flies, the neighborhood consisted mostly of densely forested bottomlands with a slight elevation to the west, the eventual site of the town of Chappell Hill. Since the eighteenth century, the Tonkawa Indians had used the area for hunting and for gathering fruits and nuts in the forested bottomlands. With the arrival of Anglo and African Americans in the 1820s, the Tonkawas' first impulse was to incorporate the newcomers into the existing trading network. And the early settlers often found it necessary to trade for food and skins to survive. One family's first harvest on New Year's Creek in 1822 yielded "scarcely . . . as much corn as we had planted," but they survived in part by tapping into existing trade connections.[22] The situation changed after Austin's 1826 war with the Karankawas in the Lower Country. Austin's men killed several Tonkawas who had intermarried with the Karankawas, setting off a series of livestock raids by the Tonkawas, which continued sporadically for more than a decade. By the 1840s, however, with the Upper Country population growing denser, more agricultural, and less dependent on Indian trade, the Tonkawas withdrew well to the north, eventually to be forced onto a reservation in 1855.

Chappell Hill proper did not emerge until 1847, when it was formed out of three very different clusters of settlers. The first of these centered on a sawmill erected in the early 1840s by three Georgians, William, Gilbert, and Terrell Jackson. Financing from a Houston merchant allowed Terrell Jackson to cultivate the bottomland near New Year's Creek with sixteen slaves as early as 1843. The same merchants sustained him through the depression of the forties by accepting land as security for supplies.[23] By 1850 Terrell Jackson owned more than one thousand acres of land, of which two hundred were "improved." That year his twenty-six enslaved laborers produced forty bales of cotton. The mill, along with the brothers' store, became a focal point for the surrounding farms.[24]

Providence Baptist Church formed the nucleus of a second forerunner to Chappell Hill. Deacon Oliver Hazard Perry Garrett came to Texas initially in 1838 for the "cheap land," but he soon returned to South Carolina to marry. In 1842 he came back to the New Year's Creek area and bought a tract, which he worked with a single slave. Although he denied being a speculator, he was a surveyor for many years and did a side business locating land for other buyers, and the area became known as a Baptist settlement. Garrett prospered as he preached, eventually becoming master of fifteen slaves.[25] The third predecessor settlement of Chappell Hill, and the one responsible for its name, was a knot of farms owned by members of the Chappell family of Tennessee. Robert Wooding Chappell came to the area in 1838 and received a grant of land from the Texas Republic in 1841. According to local legend, Chappell was one of those stock-raiser-hunters who was dismayed by the area's transformation into a plantation region and left in 1850. By then there were four Chappell households in the area.[26]

As Chappell Hill became part of the plantation world, its demography changed. A census snapshot from 1850, three years after the founding of the town, suggests a neighborhood in transition between Nestor Clay's world and that of the planters. Of the 57 resident households, 27 owned slaves, with 10 being the median holding. Only 17 households reported producing any cotton. Ten years later, the situation had changed dramatically. Slaves constituted a 68 percent majority in a population of nearly three thousand, which was concentrated in the bottomlands on the west bank of the Brazos and around New Year's Creek. The area's planters, along with a small population of tradesmen and merchants, lived for the most part in and around the town, which was situated on a ridge a few miles to the west. Sixty-three percent of Chappell Hill households owned slaves, with 39 "planters" owning 20 or more, and a median holding of 14. Cotton production had increased greatly, with 25 planters reporting more than one hundred bales. But cotton production was not universal—only 5 nonslaveholders produced cotton, probably because so many of them worked in town in the trades and professions. Elsewhere in the Upper Country nonslaveholders produced more cotton.[27]

Behind the social and economic transformation of Chappell Hill was a significant secondary migration. A comparison of the 1850 and 1860 residents reveals a startling turnover in population. Only 13 of 57 of the 1850 heads of household were still in the area in 1860, though not surprisingly, persistence was greatest among the largest slaveholders. A striking number of them also had mercantile experience of some sort, belying the popular image of the decadent, agrarian ideologue. John Washington Lockhart, an Alabama native, studied medicine at Louisville Medical College and ran a pharmacy in nearby Washington, Texas, before inheriting bondsmen and settling in Chappell Hill after 1850. Lockhart's father-in-law, Jonathan Wallis, followed a similar path. He arrived in 1848 from Noxubee County, Mississippi, with his father, his younger brother, and a sizable slave labor force. The following year, at age twenty, he opened a store, which he ran for three years before selling out to his younger brother, Joseph Wallis. Aided by his father, who divided his property among his sons in 1856, Jonathan Wallis's plantation prospered—in 1850 he owned 23 slaves and produced 8 bales of cotton; in 1860 he owned 77 slaves and produced 400 bales of cotton. Brother Joseph, still in his teens, ran the store until receiving property from his father in 1856, then began planting. In all, at least ten of the top twenty slaveholders in Chappell Hill in 1860 had some sort of mercantile or commercial experience other than planting. Some left their businesses to become planters, but others planted and engaged in commercial ventures at the same time.[28]

Before we move on to other migrations, it is worthwhile revisiting the migration of the enslaved. Although the migrations of masters and slaves were tightly connected, it is important to remember that they were not identical, largely because slaves were legally defined as chattel. For example, the source regions for slave migrants likely differed somewhat from those of their masters, with more slaves born in the Upper South, reflecting the tendency of some planters to purchase laborers on the market before heading west. And while Anglo and African American migrants both left extended kin behind, the prevalence of abroad marriage meant that African Americans were more likely to leave close family members behind. For this reason the slaves who moved with planters were often separated from loved ones and communities just as surely as if they had been sold on the market. In addition, for many slaves migration also ruptured longstanding but unofficial agreements on the nature of work and the basic conditions of life. The new environment necessarily altered both of these, as land-clearing overwhelmed more familiar crop routines. For those raised in non-cotton-growing areas, the old routines never returned and a new modus vivendi had to be negotiated. Yet for all of its trauma, migration to the Brazos also presented the enslaved with distinct opportunities. They soon learned that the Mexican government did not fully support the right to own human property, and after Texas independence in 1836, thousands of African Americans surreptitiously continued their migration deep into the heart of Mexico. We pick up their story in subsequent chapters.[29]

Speculations

From the beginning of the cotton boom in the 1790s to the coming of the Civil War in 1860, just over 1 million slaves migrated from the seaboard states into the Old Southwest. Approximately two-thirds were sold there on a highly elaborated interregional market, the major outlet of which was at New Orleans. The Crescent City in turn supplied a number of local slave markets, including Houston and Galveston, although Texas probably did not participate heavily in this trade until the 1840s. One reason was that, until 1836, the legal standing of slavery was still very much in doubt, which meant that any trader who brought in a coffle or a shipment assumed a tremendous economic risk. The proslavery constitution of 1836, which explicitly permitted the importation of slaves from the United States while outlawing the African trade, as well as the state constitution of 1845, eliminated much of this risk. A second reason for the slow start to the trade was the relatively immature state of Texas plantation society before the 1850s. With the possible exception of the Gulf Prairie and a few other Lower Country planters, few in Texas possessed the cash or credit necessary to participate heavily in the interregional slave trade. But with more slave owners establishing themselves in the state following annexation, Texas was poised to become a major buyer on the New Orleans market.[30]

Anecdotal evidence suggests that some of these purchases were arranged informally by slave owners on trips to the eastern United States. Sarah Ashley, a native of Mississippi, recalled being purchased by a Texan named Mose Davis. Ashley had been sold to a trader in Georgia, and Davis had apparently undertaken an eastern expedition to buy children for his Texas operations.[31] Another planter, Guy M. Bryan, who had come east on a similar trip, had to cancel plans to visit his friend, future president Rutherford B. Hayes, because he had "shipped some negroes to Texas before I left Baltimore," adding, "I return to complete my purchases, & if I am successful as I think I may be I shall not be able to return as I very much desire through Ohio."[32]

Most of the slaves sold to the Brazos, however, probably arrived as part of a more systematic trade carried out by the infamous "speculators." The experience of Betty Simmons, who was born in Alabama, was probably typical in many respects. As she recounted years later to a WPA interviewer, her master sold her to a trader as his financial affairs worsened. Like many slave owners, Simmons's resorted to deception in order to minimize any resistance. He told Simmons that he was going to hide her from creditors in a closed buggy, which, it turned out, belonged to a slave dealer. The man then took her to a trader's yard in Memphis, and she was placed, along with over 250 other slaves, on a steamboat to Vicksburg, Mississippi. From there they boarded a train to New Orleans and upon arrival were placed in a large holding pen. Eventually a Texan named Fortescue bought her and carried her up the Red River into East Texas.[33]

Circumstantial evidence at the local level suggests heavy involvement on the part of Brazos planters, especially during the 1850s. In Chappell Hill there were nine farmers with 10 or more slaves appearing in both the 1850 and 1860 censuses. In 1850, these nine owned a total of 148 slaves, or 16 each. In 1860, the same nine owned 355 slaves, or 39 each, a rate of increase of nearly 140 percent, far outstripping the 23 percent increase in the U.S. slave population over the same period. The increase for Gulf Prairie was also substantial; it was less spectacular, though, probably because the local plantations had already come through a period of expansion in the previous decades. There, six planters appeared in both the censuses of 1850 and 1860. In 1850, these six owned a total of 536 slaves, or 89 each; ten years later, the same six increased their aggregate holdings to 748, a rate of increase of 40 percent, well above the national rate (see table 3).[34]

Historians have struggled, with mixed results, to devise a reliable and systematic method of measuring the interregional trade. Based on a quantitative analysis of the WPA narratives, one historian found that 28.2 percent of the interviewees were brought to Texas with their masters, 5.5 percent were sold there by traders, and 62.4 percent were born in the state. Estimates based on the WPA interviews, however, are likely to understate the importance of the interregional trade to Texas. Because the interviews were conducted in the late 1930s, the interviewees had been quite young during slavery. Some were even infants. Most of their parents were almost certainly born elsewhere and of course had died before the 1930s. Another group of historians based their estimate on the sex ratios of the destination states, on the assumption that an interregional trade would have resulted in a disproportionate number of men, since male laborers were more highly valued than females. Based on this assumption, they have estimated the percentage of slaves sold on the market at between 7 and 16 percent. However, subsequent studies have called into question the assumption that the interregional trade was sex-selective, leaving those conclusions in doubt.[35]

Historian Michael Tadman has proposed a different method, based on the assumption that the interregional trade was age-selective rather than sex-selective. When planters migrated, argues Tadman, they generally took all of their slaves with them within the first year, which means that planter migrations were neither sex- nor age-selective. Using census statistics on slave ages, he calculated the increase of slaves in age cohorts over a ten-year period. The totals were then multiplied by the survival rate for each age cohort, as calculated for the South as a whole. The difference between the expected number of survivors and the actual number of slaves in a given cohort represents the approximate number of slaves imported. To separate those who migrated with their owners from those sold to the area on the market, he found the age cohort with the lowest rate of increase. If indeed the planter migrations were not age-selective but the interregional market was, this finding should represent the maximum number of slaves who came with their masters.[36]

Applying Tadman's method, we find that the interregional market was responsible for the movement of just under sixty-five hundred slaves to the area between 1850 and 1860, accounting for 59 percent of the growth in the slave population over that ten-year period. This figure is consistent with the 60 percent Tadman suggests for the entire slave-importing area as a whole. If the assumptions are correct, the interregional slave trade to the Brazos was by no means inconsequential. The trade divided and decimated thousands of families and communities, the qualitative aspects of which are historically unrecoverable. Yet, as we will see, despite the misery and disruption, slaves were capable of forming new and meaningful familial and community relationships.[37]

Osnabrück-on-the-Brazos

Mentioned frequently in the lists of goods that southern slave owners bought from their factors and merchants, amid the usual shoes, tobacco, whiskey, and silverware, were the ubiquitous "osnaburgs." The word was a corruption of the name for a coarse linen originally from the city of Osnabrück in the Kingdom of Hannover in the German Confederation. By the nineteenth century, however, factory-made cotton osnaburg, produced in a number of locales along with other cheap textiles, had supplanted the original and was now the fabric of choice for slave clothing. In one of the many small ironies of the nineteenth-century Atlantic world, a modest number of those who once wove linen osnaburgs left the area that had given its name to the fabric to start over again along the Brazos as suppliers of cotton to the very mills that had helped displace them.[38]

On the eve of the Civil War, there were nearly three thousand foreign-born residents along the Brazos, the vast majority from German-speaking Europe (see table 4). They came not only from all corners of the German Confederation—Brandenburg, Mecklenburg, Oldenburg, Hannover, Nassau, Saxony, Westphalia, Baden, Württemberg, Bavaria, and Silesia—but also from Switzerland, parts of the Austrian Empire, especially Bohemia and Moravia, and from border regions like Alsace. That said, different areas of central Europe tended to contribute population at different times, imparting a regional flavor to the migrations. In addition, individual enclaves drew disproportionately on particular regions of Germany through the process of chain migration. Most of the antebellum immigrants to the Brazos came from northwestern and western Germany, or the area west of the Elbe River and north of the Main River, especially Oldenburg, Hannover, and Westphalia, though notable contingents arrived from all regions. If the paucity of references to Roman Catholicism and the absence of a Catholic Church can be taken as evidence, then Protestantism appears to have dominated among Brazos immigrants, even among the small number of Czech-speakers.[39]

As with other Brazos residents, long-term shifts in the Atlantic economy and the global distribution of labor fueled the migration. In the case of the German-speakers, collapse of the Early Modern rural economy, which centered on a combination of small, often marginal farming and home linen production, was responsible for a mini-diaspora. Historians have described the economic system of the northwestern German states as "proto-industrial." In proto-industrial areas, a farm owner or renter would typically grow or buy flax, then the family would spin it into yarn. Women and children wove the yarn into bolts of linen cloth, which were then bought and resold by local merchants. Engagement with distant markets and merchant intermediaries eventually created a credit dependency that resulted in many cottagers' losing control over production decisions, altering the texture of domestic relations. Proto-industrialization had demographic consequences as well: since young people no longer needed to wait to inherit land to start their families, populations grew, straining the resources of rural communities.[40]

The various forms of land tenure and inheritance combined with these changes to set the pattern for German out-migration. In most northwestern states, a system of land tenure called *Grundherrschaft* prevailed. Under Grundherrschaft, peasants discharged their feudal obligations through rent payments, which by the nineteenth century were usually paid in cash rather than kind. This development in turn allowed home-produced textiles to flourish to an even greater degree than elsewhere. Moreover, property in northwestern Germany followed the tradition of *Anerbenrecht*, or entail, which meant it was inherited intact and not divided among heirs. At the start of the nineteenth century, the combination of these two legal traditions brought tremendous strain to rural northwestern Germany, as the population increased while the number of farms held steady. In the Parish of Belm, Osnabrück, the number of households jumped 57 percent between 1772 and 1833, while the number of farms increased by only 17 percent. An unprecedented number of landless people, known as *Heuerleute*, lived marginal lives as renters and wage workers. Like landowners, they too combined agricultural labor (though they worked for others rather than themselves) with home textile production. In the Southwest, a different inheritance regime created parallel problems. With a tradition of partible inheritance rather than entail, most children inherited property, so the percentage of the landless was lower, but the farms eventually became so small that families found it necessary to engage in additional activities, such as home textile production.[41]

Over the course of the nineteenth century, the strain grew too much to bear, and the entire system of rural production fell apart. The Napoleonic wars isolated producers from foreign markets. Then with the postwar dissolution of the Continental System's protections, rural textile producers found themselves exposed to increased competition within Europe. The rise of factory-produced textiles in England, Ireland, and Belgium dealt many farmer-linen-producers a blow from which they could not recover. Migration emerged as one strategy to meet the crisis, with several distinct flows. Some moved from declining rural proto-industrial areas to nearby urban industrial centers. Migration to other European countries was a second possibility. Many rural northwestern Germans had long supplemented their livelihoods with seasonal labor in the Netherlands; some of these simply made Holland a permanent home. But for many others, migration to the New World was an attractive alternative to eking out an existence in farming and home textiles. One new Texan attempted to persuade family members to join him by sarcastically repeating the refrain, "Ah, I would rather spin than go to America."[42] In the earliest phase of the nineteenth-century emigration cycle, between 1816 and 1830, Latin America vied with North America as a favored destination.[43] A large proportion of the early emigrants came from southwestern Germany. Most of these southwesterners sailed out of le Havre and other ports, but comparatively few actually came to the Brazos.

With the construction and opening of Bremerhaven on the Weser River in 1830, Bremen emerged as the principal embarkation point for central Europe, drawing on the declining proto-industrial areas of Hannover, Oldenburg,

Nassau, Waldeck, and the Münster District. The early migrants to the Brazos usually sailed first to New York or Baltimore, then to New Orleans. Later, in the 1840s, New Orleans emerged as a favored port of entry for Germans headed upriver to St. Louis. Starting in 1842, New Orleans and Bremerhaven were linked by regular service, which established a lasting and dependable link to Texas. More than 8,000 came to Texas in 1846 alone. The upheavals of 1848 spawned further migration, although the "Forty-eighters" do not appear to have been statistically significant along the Brazos, a few prominent community leaders notwithstanding. By 1857 there were between 35,000 and 40,000 German-speakers in the state, including the Texas-born children of migrants. The Brazos received only a fraction of these, but enough so that by 1860, 24 percent of the free population of Austin County and 21 percent of the free residents of Washington County were foreign born. Emblematic of the new ties between Texas and Germany, Bremen merchants began to buy southern cotton so that their ships would not return in ballast.[44]

There is an important, yet frequently overlooked, question regarding antebellum migration from Germany to the Brazos: why did these immigrants come to an area dominated by plantation slavery? Plantation agriculture had a tendency to inhibit urban development, drive up land prices, and depress both the demand for and the price of free labor. Published letters from migrants, a staple of German newspapers in these years, generally confirm this wisdom. One South Carolina migrant warned fellow Germans away from the South, writing that, in addition to the unhealthy climate, "the slave trade is very significant here because all field crops must be worked with Negroes." Most antebellum immigrants consequently avoided the South, and those who did settle there usually preferred cities or rural areas outside the plantation belt. With a classic plantation economy and a slave majority in place by 1850, the Brazos hardly furnished the opportunities sought by most European newcomers. If the modern global migrations were, as one historian has said, "a form of labor recruitment in an international labor market," then the immigrant presence along the Brazos presents a strange anomaly.[45]

One possible explanation is that the Germans settled on marginal land unsuitable for plantation agriculture. Most of the German settlements on the Brazos were indeed located in the hilly post-oak savannahs of the Upper Country, a few miles outside the bottomlands. It was not uncommon in the South for nonslaveholders to reside in the less-fertile pockets of major plantation regions. The yeomen of the South Carolina Low Country, who lived on the sand hills between the larger cotton and rice plantation areas, were but one example.[46] But while soil quality may have played a minor role in the geographical distribution of the German migrants, it cannot alone explain the location of their settlements. While a few Upper Country neighborhoods really did have poor soil, most of the land in western Austin and Washington counties was quite good, comparable to the nonbottomland plantation areas nearby and generally better than other parts of the plantation South. By the 1850s Upper Country plantations had actually escaped the confines of the bottoms and spread into the uplands onto soils of a quality similar to those in the German areas.

The explanation lies rather in the timing of the migration. The German settlements managed to flourish in the Upper Country because they were established in the Mexican era, before the dominance of the plantation economy, when land was still cheap and plentiful. Once in place, the settlements continued to attract European migrants, retaining their distinctive character while never quite acquiring the trappings of Anglo and African American plantation society, despite their proximity to it. It was the ability of the early settlements to attract newcomers even as the plantations developed nearby that accounts for the German enclaves, not the quality of the land itself. The result was a plantation region with a patchwork quilt of internal boundaries, obvious to even the least-observant traveler.

The first German settlements along the Brazos, and indeed the first in Texas, were founded in the early 1830s in the western portion of present-day Austin County. Like so many of the Anglo settlers, the early arrivals initially benefited from Mexico's borderland population policies. The petitions of Friedrich Ernst, an Oldenburger, and Charles Fordtran, a Westphalian, must have been quite welcome to Mexican officials, filed at a time when the government was trying to dilute the Anglo presence in Texas with colonies of Mexicans and Irish. Ernst received 4,428 acres of upland post-oak savannah along fertile Mill Creek, half of which he eventually transferred to Fordtran for $165. Fordtran subsequently acquired additional land by working as a surveyor for Austin.[47]

Ernst's plan was to create a German enclave in Texas by selling off small portions of his grant to migrants from the Old Country. His first step was to publicize Texas in the German press, which he did quite successfully with a "letter" that rivaled any piece of Anglo American propaganda in its determination to portray the hot, unhealthful climate of Texas as paradise on earth. "The climate resembles that of lower Italy," he related to a "friend" who quickly published his remarks. "In the summer it is of course warmer than in Germany because the sun is almost straight overhead at its peak; however, it is not hot as long as one might suppose, as a steady, fresh east wind cools the air." Most of the content of Ernst's letter appealed directly to the concerns of distressed rural areas in the German Confederation. Mexico's generous land policies, which Ernst described in detail, sounded almost too good to be true to the fast-growing population of threatened landowners and Heuerleute. The general ease of life as portrayed by Ernst must have seemed equally wonderful. Whereas German farmers had to gather fodder to feed their livestock over the winter, in Texas there was no need—horses and cattle could graze all year on the prairies and meadows. In Germany farmers had to manure their tiny, exhausted plots, but Texas lands were so rich that they required no fertilizer. The loss of common lands limited the amount of livestock one could own in Germany, but the Upper Country of the Brazos offered an unlimited range where cows calved after only eighteen months and six hogs became one hundred in a single year. And with livestock providing milk in abundance, one needed only to scoop the honey out of a hollow tree stump to recreate the biblical Canaan—or, better yet in Ernst's telling, an "Eden."[48]

In addition to the obvious appeal to material well-being, the descriptions by Ernst and subsequent writers invited readers to take part in the romantic drama of wilderness living and conquest. This desire was something Germans shared with Anglo migrants. For some time, German readers had been fascinated by the notion of a New World wilderness, a notion they consumed in works by writers ranging from James Fenimore Cooper to Germans such as Karl Postl and Friedrich Gerstäcker. Like Anglo Americans, German writers tended to equate the wilderness—*die Wildniss* in German—with hazy notions of freedom. The literal embodiment of wilderness freedom was the figure of the Indian, whose ostensibly nomadic existence prompted comparison with the more familiar Cossacks. But romantic as that sort of uncivilized freedom was, Germans, like their American neighbors, believed that the Wildniss was destined to be tamed. As one traveler in the 1830s described the area near Ernst's and Fordtran's establishments, "it is a land that only awaits Man, and through European industry, the most blessed in the known world, it will be uplifted." Another letter from the Brazos celebrated the liberating potential of wilderness simplicity, which the writer believed set the Germans apart from their Anglo neighbors. "The comforts, in the view of the Englishmen, are admittedly very few," wrote Adolf Fuchs in 1846 from his new farm in Cat Spring. "Our houses are cabins, our clothing poor, our beds hard, but our hearts are free and happy." Yet another migrant, after telling of a leisurely winter spent playing cards and smoking his pipe, told his parents simply, "We have more freedom than you do."[49]

Ernst's venture was successful. During the 1830s several dozen families settled on his league, which centered on the newly platted "town" of Industry. As might be expected, the initial migrants came from the area near Ernst's native Oldenburg, where word of mouth and his letter had circulated news of his proposed colony, but settlers from other regions eventually came as well. Despite its distinctly German character, Industry was in some ways quite un-German. The houses, while generally described as more comfortable than those of Anglo yeomen, were nevertheless of the log-cabin variety rather than of European half-timber construction, several examples of which existed in the German Hill Country settlements north of San Antonio. Unlike most central European villages or even the Hill Country towns, which clustered around a central square in an attempt to replicate the open-field system of cultivation, Industry was laid out lengthwise along what eventually became a major east-west artery, with farms dispersed across the landscape. Geographer Terry G. Jordan has suggested this was a result of two factors: cheap land held in fee-simple, and the prevalence of scattered farmsteads in northwestern Germany, the source area for so many early migrants.[50]

After 1833, two more German enclaves appeared, which, together with Industry fourteen miles to the north, made up the heart of the Brazos German settlements. The first, dubbed Cat Spring, was established near a site occupied by several Swiss migrants, Charles and Mary Amsler and Charles's brother Markus. The Amslers had sailed to New York along with thirty-two others but, reckoning their chances would be better in the West, soon sailed for New Orleans. After renting for a year in St. Louis, they returned to New Orleans, where they got word of Mexico's land-grant policy. Imagining that "each family might acquire a princely domain by occupying it and battling with the savage Indians," the Amslers promptly booked passage to Brazoria. From there they traveled by ox-cart to Industry, where they picked Ernst's cotton in exchange for shelter in a crude hut.[51] Once they became comfortable, Charles and Markus Amsler attempted to register headrights on the south fork of Mill Creek, but they lacked the necessary surveying and registration fees. At about this time, in early 1835, another party of migrants arrived with dreams of hunting, fishing, and "pioneer life," inspired directly by the Ernst letter. The party comprised the family of Louis von Roeder, a minor noble of modest circumstances from Rhineland Prussia, his sister Rosa Kleberg and her husband Robert, and Robert's brother Louis Kleberg. Possessing a "bag of Dubloons" worth eight thousand dollars, Louis von Roeder reportedly agreed to pay the surveying and registration fees for the Amslers in exchange for part of their headright. Cat Spring grew steadily, and around 1845 a twin settlement of Millheim was founded four miles away.[52]

In the 1830s and 1840s, a substantial number of Anglos still lived in and around the Germans in Industry, Cat Spring, and Millheim, but over time the towns became more distinctly German. Anglo squatters had built log cabins there in the 1820s, where they probably raised small patches of corn and ran hogs on the land. Fifteen years later, German residents found the remains of their abandoned houses and the fruit trees they had planted and pictured the now-departed squatters as half-civilized Leatherstockings, "wild, daring fellows, who often came into conflict with their neighbors or the law, and as a result were always looking for new hunting grounds."[53] Two migration streams funneled Germans into the area in the 1840s and 1850s. The first was rooted in transatlantic networks of kinship and community. As individual migrants established themselves along the Brazos, they formed linkages back to the Old Country, forging chains that channeled people, capital, and information back and forth across the ocean. Clusters of migrants from the same area, like the Oldenburgers who first responded to Ernst's letter, were the result. Charles Nagel, for example, recalled that several of his neighbors in Millheim during the 1850s, including the Dittmar, Reichardt, and Litzman families, were relatives by either blood or marriage and had known his parents in Brandenburg.[54] In addition to the transatlantic migration, a secondary migration came from the Hill Country, where poor planning by the organizers of a Germany colony in the 1840s helped invigorate the population of Industry, Cat Spring, and Millheim.

Chain migration was also responsible for the creation of a Czech-speaking enclave within the German settlements, which originated in 1851 when a letter from a Silesian Protestant minister named Ernst Bergmann reached Josef Lesikar in Landskroun on the Bohemia-Moravia border. The letter outlined transportation costs and provided enticing details on agriculture in Cat Spring, so Lesikar quickly organized a party of sixteen families. As it turned out, his wife prevailed upon him to abandon the plan, and he was lucky he did. Half of those who left Landskroun perished in squalid conditions on the Atlantic crossing. Lesikar, however, stayed in touch with a survivor and passed his letters on for publication in the *Moravske Noviny*, a regional newspaper. In 1853 he organized seventeen more families and left that October for Bremen, although not before obtaining from the local authorities a fraudulent medical exemption from military service for his son. Seven weeks later the party landed in Galveston, and fourteen days after that they arrived in Cat Spring, where they encountered remnants of the previous party of immigrants.[55]

Slaveholding, for reasons discussed fully in subsequent chapters, was minimal in the German settlements. One of the reasons, however, was a direct outgrowth of chain migration: a continual stream of new arrivals, many of whom were kin, supplied German landowners with a labor source that was not generally available to Anglos. To be sure, quite a few Germans found work on Anglo farms and plantations, in some cases as overseers. The vast majority, though, were unwilling to take their chances in an alien social environment and opted instead for the more familiar German areas. Ernst actually maintained a special quarter for newcomers, who often worked for him before buying their farms from him, while other residents rented land to new migrants. Another very common practice, and one rooted in Old World traditions, was the binding out of adolescent children to local families, known in German as *Dienst*. The sons and daughters of new arrivals could expect wages of up to $140 per year, which would go toward the purchase of a family farm. Although a few of the older residents, Ernst, Fordtran, and the Amslers among them, purchased slaves, for most German farmers the steady influx of new migrants provided a more affordable source of labor.

The experience of Franz Koy, a resident of Cat Spring, reveals some of the strategizing that lay behind a move to the Brazos. A native of Kadlup in Strelitz, Koy was a cabinetmaker by trade but, like so many others, farmed as well. Although we have no direct information on the reasons for Koy's move, it was probably in response to the usual economic stresses. He seems to have left in mid-1852, sailing from Bremen and arriving in Galveston. His wife and children he left behind, an atypical decision for most married migrants, perhaps prompted by money considerations. Although Koy was apparently not preceded by any close kin, he did meet several acquaintances from back home, suggestive of the information networks that structured so many transatlantic crossings. As a skilled tradesman, Koy found that he had "very much to do" and busied himself with the construction of two mills, a cotton gin, and a cotton press. In between jobs, he told his wife that he expected to secure farm work for as much as $1.50 per day. March 1853 found him living at Ernst Bergmann's, where he paid for his room and board by building various articles of furniture.[56]

Koy constantly kept an eye out for a good piece of land and after a few months of work was seriously considering two properties. The first, near Cat Spring, had 450 acres of "very good land," 300 of which were forested, at a cost of $400. The second consisted of only 100 acres, partly covered in oak and cedar and with a herd of roaming hogs, all for $300. Koy preferred the second, owned by a widow with one son. "She very much wants to sell," he explained to his wife, "because alone it is too difficult." At the same time, he was trying to save enough money to bring over his wife and children. His correspondence ends in June 1853 with their arrival. He did manage to buy some land, appearing on the Austin County tax rolls in 1854 with 35 acres, one cow, and a wagon to his name. Whether this was the tract he described earlier is impossible to say, but Koy's journey from Germany to the Brazos and from wage worker in a German enclave to property owner typified the immigrant experience along the Brazos.[57]

Networks rooted in German identity were vital, not only to the passage itself, but to the eventual acquisition of land. To be certain, the pattern in which older migrants employed newer ones brought occasional charges of exploitation, but in most cases it worked. Terry Jordan found that over 77 percent of German farmers owned their own land in 1850, compared with 69 percent of Anglos. In Cat Spring, 71 percent of German households owned property in 1850, and more than 80 percent did in 1860. German farmers, moreover, farmed somewhat differently from Anglo farmers. Productivity of German farms was always higher than that of Anglo-owned farms, likely a reflection on the smaller size of German farms. German farmers were also quite skillful at managing their participation in the market. Compared with Anglo yeomen, who tended to grow more corn than cotton, Germans grew more market crops, tobacco in the 1830s and 1840s and cotton in the 1850s. When compared with Anglo slave owners, the Germans grew less cotton.[58]

As a result, the German neighborhoods in the Upper Country almost invariably struck residents and travelers alike as distinct. To enter them was to cross an internal frontier. Antislavery writers like Frederick Law Olmsted praised German prosperity as a hopeful sign that Texas was not condemned to be a slave state and that cotton might be profitably grown by free labor. Many others echoed the sentiments of William Steinert, who pronounced the "cleanliness and orderliness prevailing here" to be a "feast for our eyes after we had seen so many unclean and unpleasant things."[59] Even to Anglos, who were less inclined to criticize their own settlements and more likely to view the German areas with suspicion, the settlements seemed a world apart. But they were not; only a few miles separated the German areas from the bottomland plantation districts, and as we shall see, contact between neighbors was both inevitable and frequent.

"He Remembers the Customs and Habits of His Native Land"

Only a small minority of the slaves along the Brazos were African born, but their early arrival and their concentration on Lower Country plantations gave them disproportionate visibility. As late as the early twentieth century, local whites viewed the few surviving Africans as curiosities. In about 1913, merchant James P. Underwood of Brazoria County interviewed and photographed one of them, a ninety-year-old, by then known as Ned Thompson, who at age twenty had been captured in battle. Thompson was then sold to slave traders, shipped to Cuba, and carried from there to Brazoria County, where he spent the rest of his life. According to Underwood, after seventy years Thompson still "remember[ed] the customs and habits of the wild people of his native land" and "sp[oke] the language or dialect of his tribe fluently."[60] Yet, contrary to Underwood's portrayal, the Africans who lived along the Brazos (mostly in Brazoria County) were much more than curiosities. They were an important early source of labor for the region's elite planters and, given their concentration, would certainly have had an impact on the slave communities that developed later. Many of the details surrounding the Africans who lived along the Brazos are irretrievable. Sources are scarce, largely because the trade was illegal. However, the surviving information on the timing and other aspects of the migration, combined with evidence unearthed in other contexts, allows us to advance some tentative conclusions on the migrants' place in Brazos society.

The timing of the migration is easy to determine: virtually all of the Africans who lived along the Brazos arrived from Cuba between 1833 and 1840 on the outside, with the vast majority coming in 1835–1837 (see table 5).[61] Virtually all of the 123 Africans listed in the 1870 Brazoria County census were fifty or older, which is consonant with the rather young demographic profile of the Cuban trade. In addition, of the eleven known or suspected slaving voyages to the Brazos, four allegedly took place before 1835 and none occurred after 1837. Anywhere from 200 to 500 Africans were imported in 1835–1836 by a single trader, the confidence man Monroe Edwards. These years also coincide precisely with the Texas Revolution, suggesting that traders used the political chaos to provide cover for their activities. Finally, the period also coincides with high cotton prices and high slave prices in the United States. The up-and-coming, cash-poor planters of the Brazos were undoubtedly attracted by the lower prices for the Africans, not merely because they were African, but because direct importation meant that local buyers did not have to compete with wealthier Mississippi and Louisiana planters in the New Orleans market.[62]

Precisely how many Africans were imported has been a matter of speculation since the nineteenth century. Writing in 1902, Eugene C. Barker estimated the total volume of the African slave trade to Texas at 600 to 700 and under no circumstances more than 1,000. More recent scholarship has generally not contested Barker's figures. There is, however, reason to believe that 600 to 700 represents something closer to a minimum, rather than a maximum. The total number of slaves reported on the *known* slaving voyages to Texas is between 600 and 800 (see table 5), but given the illegal nature of the trade, it is entirely possible that other ships managed to land undetected. In a speech to the Congress of the Republic of Texas, President Sam Houston reported that "thousands" of Africans had been brought to Texas from Cuba. His estimate was probably not well researched, but the trade was large enough to attract the attention of the U.S. consul in Havana, who as of February 1836 estimated total slave exports to Texas at 1,000. In an 1843 report on slavery and the slave trade in Texas, the British consul at Galveston listed five separate voyages and "ascertained" the total imports to be 504, though 200 of these were apparently landed on the Sabine River in East Texas, well away from the Brazos. Moreover, Africans remained visible in Brazoria County into the postbellum era. While registering voters in 1867, a Freedmen's Bureau official reported finding 500 African-born men in the county. Finally, the federal census taken in 1870, notorious for its tendency to undercount, listed 123 African-born men and women in Brazoria County. This number is certainly lower than other estimates, but by this time almost all of the Africans were between fifty and seventy years old, suggesting an 1830s population several times as large. It seems reasonable, then, to estimate the African-born population of the Brazos at between 600 and 1,000, possibly even higher.[63]

With no surviving account books, it is difficult to reconstruct the economics of the trade, but a rough estimate suggests that the endeavor was not as profitable as the transatlantic trade. The average price for a *bozal* (the Spanish term for African-born slave) on the African coast in 1831–35 was US$37.50. The average price in Cuba during the same period was US$281.70, or a gross yield of more than 600 percent, but deductions for factorage in Africa, transoceanic shipping, and distribution in Cuba trimmed profits to 140 percent per slave. The mean slave price in Texas during 1837 (the first year with reliable data) was US$568. At those prices, traders realized a gross profit of 100 percent on slaves sold from Cuba to Texas, but transportation costs, though certainly lower than for the transatlantic slave trade, would have cut significantly into the profits. In addition, the desire to avoid detection meant that Brazos traders often employed smaller vessels (or carried small cargoes of forty or fewer), which would have increased marginal costs. Given the economic costs and legal risks, it seems that even in the "Flush Times," pure speculation in African slaves probably did not pay well, unless the trader's intent was dishonest from the start, as was the case with Monroe Edwards. Most of the Brazos traders were probably attracted as much by the prospect of using the slaves on their own plantations (and thus paying for their transportation over time through their labor) as by the opportunity for speculative profit.[64]

Evidence on the origins of the Africans who lived along the Brazos is extremely sparse, and where it exists at all, it is indirect. The most potentially valuable piece of evidence that might shed light on the matter, a "vocabulary" of African words and their English equivalents taken down by William Fairfax Gray during an encounter with two men and two women during the Runaway Scrape of 1836, has apparently not survived.[65] Other linguistic evidence, specifically the names of the Africans that have survived in various documents, is extremely problematic and must be handled with caution. While the origin of some names is reasonably clear, quirks in transcription and similarities across ethnic and linguistic groups within Africa make specific conclusions difficult. Those names that can be identified suggest a mixture of origins that more or less conforms to the profile of the slave trade to Cuba during the early 1830s. Our reconstruction of the journey from Africa to Cuba to the Brazos, then, begins with an examination of the Cuban trade.

Strictly speaking, the transportation of slaves from Africa to Cuba had been illegal since 1820, the result of an agreement concluded between Great Britain and Spain in 1817. The trade nevertheless continued owing to loopholes in the suppression treaties. Penalties for violations were weak and applied only to those involved in actually transporting captives. Adjudication procedures for intercepted ships were compromised by the byzantine mechanics of the Mixed Commissions, in which the fate of a vessel and its cargo was often determined by a drawing of lots. To make matters worse, as late as 1830 the suppression treaty between Brazil and Britain actually permitted slave trading below the equator, prompting traders of many nations to continue their activities under flags of convenience. Further compounding the weakness of international agreements on the slave trade was the failure of Britain (and eventually of the United States and France) to allocate sufficient naval resources to interdiction. Finally, on the demand side in Cuba itself, officials from the captains-general down accepted bribes for looking the other way and partook of the fruits of illegal ventures. In Cuba, even those slaves who were found to be illegally held were bound out to toil on public works, including labor for crown officials and for private estates, for periods ranging up to eight years. Known as *emancipados*, these unfortunates were often reenslaved once they disappeared from public view.[66]

During 1831–35, the five years overlapping the introduction of Africans into Texas, Cuba imported more than 85,000 slaves from across the Atlantic.[67] Among those whose origin is known, a single stretch of coastline, the Bight of Benin, saw the departure of the largest share, about 42 percent. The regions from which the next-largest percentages embarked, West Central Africa and Sierra Leone, contributed almost equally to the total, 22 and 21 percent, respectively. A fourth region, the Bight of Biafra, was responsible for only 15 percent of the Africans imported into Cuba in these years.[68] Whether the Africans who were transported to the Brazos conformed proportionally to this distribution is impossible to say for certain, but there is reason to think they did. Although the particular backgrounds of the Africans entering the diaspora at any given moment were determined largely by local contingencies—wars, droughts, politics, and economics—and were anything but random, the trade from Cuba to the Brazos was a secondary one that likely represented a more arbitrary draft on the population of Africans already present on Cuban ingenios. The only account of the purchase of Africans for transport to Texas (secondhand, unfortunately) underlines its randomness. According to one Brazos resident who claimed acquaintance with Texas slave traders during the 1830s, Brazoria planters Ben Fort Smith and two of the McNeel brothers visited a plantation outside Havana and selected fifteen bozales from among two hundred present, with no apparent attention to ethnic background. Assuming the migration followed this general outline, it would likely have resulted in a near replication of the distribution on the island at large.[69] To understand the backgrounds of the Africans, then, we must turn to the source regions on the continent itself.

The Bight of Benin was the single largest contributor to the African population of Cuba in 1831–35, with a share of over 40 percent. Known during the era as the Slave Coast, the Bight of Benin was the second-ranking region of embarkation throughout the entire era of the slave trade. Slaves in this region came from numerous linguistic and cultural backgrounds, including Fon- and Yoruba-speakers, and were enslaved via several different avenues, ranging from imperial conflict, to judicial condemnations, to raids hundreds of miles inland. War was the catalyst for most enslavement during the 1820s and 1830s. Beginning in the late eighteenth century, the most powerful state in the region, the Yoruba-speaking Oyo Empire, began to weaken. Whereas Oyo had previously relied on sources outside the empire to supply Atlantic traders, it increasingly harvested its own population, depopulating itself as it stirred tremendous animosity. The final blow to Oyo's power came in the early nineteenth century with an Islamic holy war, a jihad waged both within the kingdom by marginalized Muslim Yoruba-speakers, and without, by Muslim Fulbe [Fulani] herders who had migrated into the northern marches of the empire to energize their coreligionists among the local Hausa-speakers. A new Islamic state to the north, the Sokoto Caliphate, sponsored further jihads, precipitating the final collapse of Oyo in 1836. As Oyo declined, its residents were increasingly subject to capture and sale, leading to a surge in the number of Yoruba-speakers in the New World, particularly in Bahia and Cuba.[70]

Surviving evidence tends to confirm the presence of Yoruba-speakers along the Brazos. Among the names of Africans recorded in estate inventories and the census of 1870 are several of probable Yoruba origin, names such as Ojo, Ajie, and Ega. A second piece of evidence comes from a legal battle waged during the 1840s. The original plaintiff, Thomas Toby, sued Lower Country planter David Randon in the early 1840s for payment on several notes. Randon, already in debt to Galveston merchant Thomas McKinney, had surrendered several African slaves to McKinney. When Toby tried to attach the slaves, Randon's lawyers argued that they were not legitimate slave property, having been imported illegally, and therefore not subject to attachment. The court then took testimony on whether the Africans had been legitimately enslaved while in Africa, and Randon brought in an "expert" witness to testify specifically whether slavery existed among the "Lucame tribe."[71] *Lucumí* was a common Cuban term for Yoruba-speakers, and the court's interest in the question suggests that the slaves at one point had been identified as such. Yet while Oyo was probably the largest contributor of captives in these years, it was not the only source in the region. The catchment area for the Bight of Benin ranged from the Volta River in the west to the Niger in the east, with Hausas, Nupes, and Igbos drawn into the trade along with Yoruba-speakers. As historian David Eltis has written, it also overlapped with that of the Kingdom of Dahomey, bringing a number of Fon-speakers into the diaspora.[72]

Slaves leaving from the Bight of Benin in the 1830s would likely have embarked from Lagos or Ouidah, with smaller numbers from Godomey, Badagry, and Grand and Little Popo. Sheltered by a lagoon that made it attractive to smugglers, Lagos was an independent city-state that displaced Ouidah as the region's most prolific slaving port in the early nineteenth century. Captives entered the city from the north and were housed in buildings of various sizes that served as barracoons. There they often stayed for months, segregated by sex and chained to each other at the neck. Samuel Ajayi Crowther, enslaved in 1822, housed in Lagos, then liberated by a British patroller, recalled that the chains choked and bruised the adult men. At night the men would try to create slack by drawing the chains toward their bodies, in the process nearly suffocating the boys chained to them. Joseph Wright, who had a similar experience in 1827, recalled the danger bulking posed in case of fire. He witnessed the deaths of fifty or so people who were unable to escape a burning building because of their chains.[73] Those shipped from Ouidah and nearby ports such as Godomey had to contend with a different set of hazards. Because the waters were too shallow for oceangoing vessels, slaves had to pass through deadly surf to be placed onboard ship. Surviving the trip required a great deal of experience with local conditions, and traffickers were forced to hire boatmen to transport slaves for them. The city's exposure to British patrollers also made it risky for slavers to embark from there. Instead, most sent slaves to other places on the lagoon and embarked from there.[74]

If a ship managed to evade the West Africa Squadron, the journey from the Bight of Benin to Cuba lasted on average fifty-two days. The mean mortality rate for the Middle Passage between the Bight of Benin and Cuba in the early 1830s was almost 11 percent, but there was significant variation. In general, longer voyages meant more deaths, often caused by gastrointestinal ailments like diarrhea and dysentery.[75] Although slave ships sometimes sailed straight into Havana, most slavers thought it wiser to land nearby, outside the gaze of British diplomats and commissioners. According to depositions given in the famous *Amistad* case of 1839, the captives made initial landfall at a village one day from Havana. Five days later their captors took them to a second town on the outskirts, where they stayed five more days before being taken into the city under cover of darkness. Those who eventually made their way to the Brazos were apparently taken to ingenios not far from the city, where they were eventually purchased by Texas traders.[76]

If we are correct in assuming that the distribution of Africans along the Brazos mirrored the distribution of bozales in Cuba, the second-largest contingent of Africans, perhaps between one hundred and two hundred, embarked from ports along the West Central African coastline. A few of the slave names from inventories, such as the adapted place names "Congo" and "Bengola," support the notion of a Central African presence. This should come as no surprise—Central Africa was the leading source region for the transatlantic slave trade, accounting for almost half of all captives shipped to the Americas. The years 1831–35 were a time of transition for Central African traders. In 1830, British pressure forced Portugal and Brazil to cease trading through Luanda, a longtime shipping and financial center. In response traders shifted their operations northward to the Zaire (or Congo) River and the area above that, known as the Loango Coast. The vastness of the region's hinterland meant that the captives who came through these areas came from a variety of backgrounds. By the nineteenth century, trade routes stretched hundreds of miles into the continent, encompassing a wide variety of local communities and polities. A minority consisting of Mayombes probably came from areas near the coast. A larger number still probably consisted of Kikongo-speakers from the area near the Malebo Pool, about two hundred miles up the Zaire.[77] Common language, however, did not necessarily translate into common identity. According to historian Joseph C. Miller, local identities were often rooted more in kinship and agricultural ways rather than in language or dynastic politics. Identity formation in the West Central African diaspora was therefore not automatic; rather, the enslaved had to search out common ground to create new connections, and they often succeeded.[78] Yet because of this situation, and because their numbers were likely smaller than the numbers of Yoruba-speakers, it is reasonable to assume that West Central Africans were likely less visible and less cohesive and exerted less influence on Brazos slave society.

Sierra Leone and the Bight of Biafra accounted for 21 percent and 15 percent, respectively, of slaves entering Cuba in 1831–35. It was quite ironic in the case of the Sierra Leone, since Freetown was the site of both the Mixed Commission and a colony of Africans liberated from the slave ships by the British Navy. The region had not been heavily slaved in previous centuries, but naval suppression made it more attractive during the nineteenth. A coastline punctuated by rivers and broken up by mangrove swamps made Gallinas, Sherbro, Rio Pongo, and Bissau into prominent centers of the trade. Gallinas, the largest of them, drew on a hinterland that contained Mende-, Fulbe-, and Bambara-speakers, a few of whom likely ended up on the Brazos. The Bight of Biafra, which lay to the east of the Bight of Benin and the Niger River delta, had been a major source of labor for the British colonies during the eighteenth century. Here the commerce in humans was largely in the hands of Efik traders, organized into associations like the Ekpe, who exchanged Igbo and Ibibio captives taken in raids for the usual assortment of textiles and other trade goods.[79] Given the diversity of slaves embarking from Sierra Leone and the small numbers from the Bight of Biafra, it would have been extremely challenging for captives from either coast to reconstitute their specific, local identities along the Brazos. More likely, they found it necessary to build bridges with Yorubas and other Africans, as well as with creole slaves.

Once on the Brazos, various factors combined to maximize the visibility and influence of African-born slaves, at least for a time. Most important among these was their concentration in Brazoria County. At the time of the Africans' arrival, the slave population of Brazoria County was about 1,100. If we estimate the African population conservatively at about 500 in 1837, Africans would have constituted about half of the entire slave population. Continued migration from the United States would have eroded this percentage immediately, but for a moment in the 1830s the newcomers constituted a significant share of the laborers in Texas's most productive plantation county. It also appears that after some temporary movement during the Runaway Scrape of 1836, an overwhelming majority settled on just a few Lower Country plantations. Of the approximately fifty plantations in the county, there is documented evidence of at least one African-born slave on fourteen of them. Moreover, African-born slaves seem to have been geographically concentrated in just a few areas: Monroe Edwards's Chenango Plantation on Oyster Creek in the northern part of the county; the Mims plantation on the west bank of the San Bernard River; and most of all, the Gulf Prairie neighborhood, where there is evidence of Africans on all of the plantations except the Westall place and Joel Bryan's Durazno (see map 2). Lastly, accounts from the 1830s of groups of slaves conversing in at least two different African languages leave no doubt that many of the Africans could communicate with each other, which in turn raises the likelihood that African speech communities existed in Brazoria County. Together these factors would have maximized—though not in all cases guaranteed—the opportunities to form families and preserve knowledge and lifeways from Africa, as we shall see in chapter 2.[80]

These migrations were, each in its own different way, highly structured, never random. To make matters more complicated, the successive border and settlement policies of Mexico, the Texas Republic, and the United States all affected the dispersion of migrants, whether it involved an invitation to settlers in the 1820s, an effort to bar some of them in 1830s, or a redrawing of the border to put more distance between Mexico and the plantation regions of Texas. As a result, the Brazos in no way resembled the "melting pot" of popular American myth; instead, it more closely approximated a mosaic of seemingly distinct neighborhoods. At the same time, however, we should keep in mind that difference and diversity are historically and culturally constructed, never absolute or essential. While it may seem obvious that the Germans were different from their Anglo neighbors, or that Africans were different from American-born slaves, or even that "white" residents were different from "black" residents, we miss a great deal by simply assuming these things. What E. P. Thompson observed of class applies to all types of difference: it happens when some, "as a result of common experiences (inherited or shared), feel and articulate the identity of their interests as between themselves, and as against other men whose interests are different from (and usually opposed to) theirs." Although conflict, between men and women, masters and slaves, and neighbors, emerged almost immediately, inscribing palpable boundaries on the bodies and landscapes of the Brazos, residents were simultaneously linked to each other by new, improvised border languages. It is to the creation of those languages that we now turn.[81]

Chapter 2

HUSBANDS AND WIVES

After processing his first wave of land claims in 1824, Stephen F. Austin received a curious letter from one of his deputies, surveyor Seth Ingram. A settler named Thomas J. Tone had filed for a league of land (4,428 acres) with a partner, Thomas Jamison. Tone believed himself entitled to more than the half league he would receive as Jamison's partner, and he protested to Ingram that he was "a Man of a family and ought to be entitled to the same priviliges as other Men of families." In fact, Tone's request was not an outrageous one: Austin did occasionally "couple" single men for the purposes of sharing a league, but Tone was not single. For his part, surveyor Ingram endorsed Tone's position, telling Austin that "Mr Tone has suffered in this country with the rest of us, and is what a Man ought to be," adding that "it appears that some disturbance is likely to arise in consequence of the division of the Leagues, when two Men is coupled together and Making choice in different Leagues, and threatening if they cannot have their land so and so they will go to the *Guadaloupe* [River, southwest of the Austin Colony]."[1]

As the wording of Ingram and Tone's protest demonstrates, the concept of family was important to Anglo plantation society for ideological as well as material reasons. To deny Tone a full league of land not only would have left him with a smaller share of the productive resources; it threatened to make him something less than a man. The society from which settlers like Tone hailed, rooted as it was in slavery, was a highly gendered one in which mastery of the household (the primary locus of production) and family (the dominant metaphor of power) played an integral part. The ideological symmetry between mastery over one's slaves and dominion over one's wife and children was one of the things that held southern society together, giving slaveholders and nonslaveholders a common interest and language. But marriage was important for other reasons as well, specifically in connection with evolving concepts of citizenship. Family was frequently imagined as the fundamental constituent part of the state, the element that made possible the independence of the ideal citizen. Conversely, the privilege of marriage helped construct citizenship itself. Those who did not marry in a fashion recognizable to Anglo and Mexican officials, specifically indios bárbaros like the Karankawas and the Tonkawas, were excluded from the body politic. Similarly, the denial of legal marriage to the enslaved was one of many practices that helped to define the black-white racial binary, another determinant of citizenship. Of course, actual family and household relations seldom conformed precisely to the ideal as expressed in religious and political literature, but the ideal persisted.[2]

Marriage was central to Mexican society as well, and for the most part it functioned similarly despite certain religious, political, and legal differences.[3] Those in charge of policy in the northern marches of the Mexican Empire and republic generally viewed marriage as a token of civilization and membership in the nation and viewed household formation pursuant to agriculture as the most promising path to economic development. Marriage and family became cornerstones of Mexico's policies in its northeastern borderlands. Yet, though Mexican officials and Anglo settlers regarded marriage in similar terms, for a variety of reasons Texas's borderland situation prevented a clean transplanting of southern gender ways, at least until after independence in 1836. Up to that time, and even beyond, the failure to place marriage on a firm footing helped weaken Anglo plantation society as a whole, for example, by failing in many instances to provide for the solemnization of marriage. In addition, the civil code on which Mexican law was based allowed married women to maintain separate property, a provision that some Anglo women leveraged to advantage in their marriages. After independence, Texas adopted common law but retained the property provisions of the civil code, placing matrimony on a firmer footing, but even as the legal status of marriage stabilized, the legacy of Mexico's border policies continued to complicate efforts to replicate Anglo marriage ways. The sheer diversity of the Brazos borderland fostered multiple understandings of marriage. Some understandings, such as those of Africans and African Americans, posed little threat because their marriages enjoyed no legal status. Others, such as those of the Germans, enjoyed legitimacy but lacked the linguistic symmetry with the master-slave relationship, subtly weakening one of the pillars of Brazos slave society.

Couplings

Stephen F. Austin and the Mexican authorities who worked with him viewed family as the institution that would civilize the Texas wilderness. As early as 1820, in Moses Austin's first petition to the Spanish Crown for colonization privileges, it was self-evident on all sides that "tres cientas familias" should be the fundamental building blocks of a civilized Texas. In 1823, when the imperial government of Agustín de Iturbide issued the first general colonization law, family was again the presumed unit of settlement. More than that, the law also envisioned marriage as the path to Mexican citizenship, declaring all foreigners to be automatically naturalized as long as they supported themselves for three years and were married. It also granted immediate citizenship to the spouses of Mexican citizens. Furthermore, all settlers would receive a generous grant of land—a *labor* for those who intended to raise livestock and a *sitio* for those who intended to farm—as long as they professed Roman Catholicism, presented a certificate of good character, and agreed to abide by the laws of Mexico. A revised colonization law issued by the state of Coahuila y Texas in 1825 explicitly differentiated between single and married men, awarding unmarried men one-quarter the land that married men received, with the difference claimable upon marriage. In an apparent effort to prevent the wholesale Anglicization of Texas, the law also awarded a 25 percent land bonus to any foreigner who married a Mexican citizen.[4]

An unstated assumption in all colonization legislation was that "families" would be headed by men. In practice the overwhelming majority of families were, but among Austin's initial three hundred grantees were at least nine women, and an 1826 census reported that eleven out of the 290 households present in the colony were headed by women. More striking, however, is the number of men who, like Tone and Jamison, chose to form partnerships in order to claim larger amounts of land. Twenty-seven of Austin's "Old Three Hundred" households consisted of all-male partnerships, including four "couplings" of three. Austin and the land commissioner for Coahuila y Texas, Baron de Bastrop, accepted such partnerships and awarded them a full allotment of land. But unlike married heads of households, these men would presumably have to divide the grant between them if each wanted his own freehold, and as we saw in Tone's petition, the policy did generate concern for what "a Man ought to be." Mexican policy, then, influenced household formation on the Brazos in contradictory ways. On the one hand, it bestowed clear material rewards on "families"; on the other hand, it tolerated households that diverged from the patriarchal ideal that nourished seaboard slave society.[5]

Early Brazos households were structured differently from those on the southern seaboard of the United States. Whereas the latter tended to be of the simple, or nuclear, variety, Brazos households took a variety of forms. During the earliest years, labor needs were responsible for most of the differences. From the start, many Anglo migrants hoped to meet these needs with enslaved labor, and while many of the earliest Brazos settlers owned slaves—about 35 percent in 1826—a majority did not, which meant that they would have to secure labor elsewhere. Larger households met their needs by enlisting the aid of family members, and it was not uncommon for households to pool labor by forming alliances, most often along kinship lines, leading to a surfeit of nonnuclear households. Other households acquired extra labor by hiring what the census categorized as "servants," though it is doubtful that many signed formal indentures to that effect. Noah Smithwick's experience demonstrates how informal these relationships could be. Arriving in 1825 and professing no interest in land, he found work as a blacksmith at John McNeel's Gulf Prairie plantation in exchange for shelter and sustenance during his periodic bouts with the fever. Other landowners secured labor by offering newcomers the chance to lease a section of their domain or to keep a fixed amount of the livestock increase in exchange for their labor. White servants were even more common than slaves in early Brazos households, appearing in approximately 40 percent of those listed in the 1826 census.[6]

The demographic structure of the early Brazos settlers also influenced household composition. Like most colonies in the initial stages of settlement, the Brazos and the Austin Colony had a male-heavy sex ratio. Precisely how skewed is impossible to say because Austin did not record the sex of servants or slaves, but if we assume that nearly all of the servants were men, there would have been more than two white men for every white woman. Because the migration of slaves was involuntary, the sex ratio was likely not as skewed among them, which likely meant that a disproportionate number of women were African American. Not surprisingly, quite a few Anglo men had long-term sexual and even emotional relationships with women who were enslaved to them. One of these was Columbus R. Patton, who came to Brazoria County in 1834. The owner of a sugar plantation with more than fifty slaves, Patton had a long-term affair with a slave named Rachel. According to several of the neighbors, Rachel acted as the plantation mistress, issuing directions to the overseer, including orders to whip other slaves, and donned ladies' finery to sit in the white section in church. Although it is clear that such relationships did not meet with community approval, they were easier to maintain in Mexican Texas than they were in the United States or, for that matter, after independence in 1836, because of the unbalanced sex ratio and the tolerant stance of Mexican law.[7]

A final hindrance to the replication of seaboard marriage was that Mexico did not officially tolerate Protestantism, which meant that the Austin Colony contained no actual churches and relatively few ministers of the gospel. Anglo colonists consequently found it very difficult to solemnize their marriages. Most couples who wished to marry entered into a contract in which each partner was bound to the other, with violations punished by forfeiture of a previously posted bond. The couple then swore to solemnize the marriage at the first opportunity, either upon the next visit by a Catholic priest or by a Protestant minister on a future visit by the couple to the United States. In the meantime, however, husband and wife lived in a liminal state between marriage and nonmarriage, leaving the duties of husband and wife undefined. In spite of the chance of forfeiture of the bond, women could escape a "marriage" of this kind with relative ease. One Anglo settler credited the system with "combining in itself all the essential features of both marriage and divorce," since if the couple "changed their minds before the priest got around, they had only to go together to the alcalde [magistrate] and demand the bond, which they tore to pieces and were free again." Male colonists saw de facto divorce on such easy terms as a real problem, as Austin highlighted in a letter to a priest at Bexar after encountering two settlers whose wives had abandoned them. "They ask," reported Austin, "since they are farmers who work all day in the fields, to marry again." A resolution of the situation was important because of "the necessity of having a companion who takes care of a man's house and interests."[8]

For the few free women who lived along the Brazos in the 1820s and 1830s, the unbalanced sex ratio afforded them more power than they might have wielded in the United States. The scarcity of free women conferred an advantage in negotiations with potential marriage partners, as Ann Raney learned. A well-bred Englishwoman who arrived in the 1830s, she attracted considerable attention from the young men of Brazoria, including Gulf Prairie's Sterling McNeel. During their courtship, it came out that Sterling's father, John McNeel, opposed his marriage and threatened to disinherit him. Rather than risk his father's ire, Sterling decided to play for time with Ann as he waited for his ailing father to die. As Sterling's visits grew less frequent, a handsome bachelor named John Thomas caught Ann's eye, and after friends arranged a meeting, she married him instead. For marriageable women, it was a strong seller's market. (As it turned out, Ann made the wrong choice, at least from an economic perspective. Her husband, John Thomas, ran up large debts and abandoned her during the war of 1836, but Sterling McNeel became one of the richest planters in Texas.)[9]

As the area grew more populous and settlement more dense, the sex ratio evened out and Anglo households increasingly resembled those of the seaboard South. The percentage of unmarried men and all-male households decreased as more Anglo women moved into the province and more nuclear households were established. Still, some diversity of form remained. Planter-dominated regions like Gulf Prairie had proportionately fewer nuclear households due to their reliance on overseers and skilled wage laborers, most of whom were single men who either lived in the same house as the planter or in a separate building with other wage hands. More than 60 percent of Gulf Prairie households contained nonfamily members in 1850. Areas with few or no plantations had proportionately more nuclear households. Before the growth of the plantations in 1850, at least 84 percent of Chappell Hill households contained no members who were not blood relatives. Ten years later, as local farms were transformed into plantations, the proportion of households consisting solely of family and kin dropped to about 63 percent (see table 6).[10]

As the sex ratio among whites balanced out, long-term interracial liaisons met with increased community disapproval. A case in point was the relationship between Rachel and Columbus R. Patton, discussed earlier. While Rachel's assertiveness in church and on the plantation was apparently tolerated by the community well into the 1840s, it met suddenly with determined resistance around 1850 when heirs brought a non compos mentis suit against Patton. Central to the suit was the testimony of several neighbors and former residents of the plantation, many of whom emphasized Rachel's domestic power to highlight Patton's insanity, and the heirs eventually won the case. Patton does appear to have been incapacitated, but it is nevertheless significant that the suit came at a moment when Brazos households more closely resembled their seaboard counterparts and open interracial relationships had become less common. That neighbors considered an interracial "marriage" to be evidence of insanity by 1850 suggests a will to police racial and sexual boundaries that was simply not present twenty years earlier.[11]

Accompanying these demographic changes were the inevitable political and legal changes that arrived with independence in 1836. The disestablishment of Roman Catholicism and the appearance of Protestant ministers made it easier to formalize unions. Texas legislators and courts systematically began to address perceived deficiencies in the marital codes. One of their main concerns was the practice of bond marriage, which sowed confusion regarding the legitimacy of children and the succession of estates and prompted concerns about bigamy in the event that one partner remarried. Congress and the courts generally did not penalize residents for past irregularities; and with a series of laws passed in the 1830s and 1840s, they created procedures to formalize makeshift unions. In 1841 the legislature acted to legitimate the children of all irregular unions. Another law allowed illegitimate children to inherit property from their mothers. Lastly, in a decision dating to 1846, the Texas Supreme Court ameliorated the plight of women duped into bigamous relationships by legitimating the offspring of the union and granting the wife the right to community property.[12]

Although the Republic of Texas formally replaced the civil code with common law in 1840, it specifically left in place Spanish rules on married women's property. Spanish and Mexican law allowed married women to keep in their own names all property acquired before entering the marriage. Only in special instances, however, were they allowed control of it. Wives furthermore were permitted an equal share in all property acquired during the time of the marriage, though once again they were allowed to control it only under certain circumstances. After 1836, however, unfamiliarity with Spanish codes and confusion over which were still in force led to a series of inconsistent rulings on the matter. After experimenting with several confusing statutes concerning dower rights and equity law, Texas passed a comprehensive married women's property law in 1841, requiring that wives give permission for the sale of any separate property but continuing to allow the husband control of it. Further legislation in 1846 required a wife's consent for the alienation of a homestead. Laws like these were not inspired by any desire to instill gender equality. Like numerous laws in the United States at the time, they sought to protect a woman's property from profligate husbands and poor investments. They were not revolutionary, but rather sought to protect the integrity of the household without changing its nature.[13]

For the most part, these laws achieved the goal of rationalizing the law of marriage without making radical changes. However, in isolated instances, women occasionally asserted control of property over and above what the law envisioned. In 1837, Emily Perry of Gulf Prairie, as she was about to leave on a trip to the United States, formally authorized her husband, James F. Perry, to dispose of her inheritance as he saw fit. But her concession came with detailed instructions on how to handle the estate. She directed Perry not to accept questionable currency and added, "If you have an oppertunity of buying any Negroes for Land, I would advise you to do it."[14] After returning from her trip, Emily Perry continued to manage her estate through her husband, an arrangement that left him feeling vaguely inadequate and defensive about his position. In 1851, Emily drew up a will ceding a portion of her considerable holdings to her sons from a previous marriage. James expressed his dismay, which led to a serious argument. Protesting later that he did not intend to usurp his stepsons' inheritance, he told his wife, "My object was not to ask you to give me one Dime nor one mill I never wanted it nor would not have it. . . . but I thought I had an equitable right." He then pointed out—inaccurately, since the land for his plantation had been a gift from Stephen F. Austin, Emily's brother—that he had made his fortune himself, insisting, "I live . . . without being dependant any person on earth . . . whatever has been made here & in Missouri has been made out of my means not a dollar of yours has ever been used for my benefit."[15]

Despite all the attention paid to the form and legal status of the household, Anglo settlers seldom referred to it by name. Instead, they spoke of "family," a far more elastic term whose penumbra extended well beyond the productive unit denoted by "household." In correspondence, newspaper articles, and conversation, Anglos often used the term interchangeably with household, as when James F. Perry told his son that, "Our whole family white and black have been sick." *Family* also referred to the immediate reproductive unit and its legitimate offspring, ideally, if not actually, distinct from slaves. Thomas

Affleck, a Washington County planter, made the distinction when he spoke of moving "my family & people" to Texas.[16] Defined as private in culture and law, the language of family could act as a bridge to the public sphere. One of the clumsier invocations of family to secure political favor came from Brazoria County resident Thomas Masterson, who based his application to become wreck master of the Brazos River in part on his membership in a family. In a petition to the governor, his supporters described him as being "for fifteen years a citizen of Texas, . . . the head of a large family, and . . . fully competent to discharge the duties of the office."[17]

But above all, family signified an intimate, organic, community of affection. In their communications with one another, family members employed a rich emotional vocabulary. Full participants in the nineteenth-century shift toward companionate marriage and romanticism, Anglo couples and their children spoke a language of passion. Typical was this fulsome declaration of love offered by John Washington Lockhart of Chappell Hill to his wife, Sallie:

> [Before marriage] we were strangers, now we are intimate friends, busom companions, one in sentiment thought idea and affection, and as St Paul says we have become as one flesh. Thank God for the union. My sweat wife at that time I was a mere wreck, liable to be to sea and stranded on any beach, A ship without a Ruder or Pilot, A poor miserable man I am now a different man. I have one to love one to adore, How thankful I feel to you for your kindness, in thus saveing me, you came as a ministering Angel.[18]

Although Lockhart portrayed marriage as the coming together of "busom companions," other speakers made it clear that family members were not equal: husbands were clearly expected to lead their families. Evangelical religion, which furnished a vocabulary of affection to many couples while endorsing male dominance, had not always been so patriarchal or family-friendly. In the early nineteenth century, women assumed a central role in many evangelical congregations, almost preaching in some instances. In many families where the husband showed little interest in religion, mothers took charge of their children's spiritual education, a role that had once been the exclusive right and duty of the father. Early evangelicals also exhorted their members to place spiritual families before their worldly families, which frequently led to conflict between evangelical and nonevangelical spouses. By 1830, however, most evangelical denominations had moderated their earlier radical stances and reached an accommodation with their patriarchally inclined southern congregants. Love remained central in evangelical marriages, but in exchange for affection and protection, women were expected to subordinate their desires to those of their husbands. As Lizzie Scott Neblett wrote to her sister on the occasion of a friend's marriage, "Mutual forbearance towards each others faults and concession is the strong hold of domestic happiness, and the wife should always remember that, she can concede with more grace, than the husband, consequently should practice that virtue more than the husband." Love, for Rebecca Pilsbury, implied being grateful for the "kindness and support which my husband has ever most indulgently extended to me." The editors of the *Texas Baptist* in Grimes County linked affection and mastery in an article on the benefits of family prayer sessions. "[Prayer] unites the members of the family together, gives the head of it a feeling of tenderness and affection for those under him, and makes them esteem and love him," they explained, adding that "some masters are, on this account, almost considered as angels in their families." For women, affection clearly implied obligation, forbearance, and submission. The affection within companionate marriage, far from being inimical to household hierarchy, was its mainspring.[19]

While many wives partook of this cult of affection, not all Brazos families lived up to its harmonious ideal. The relationship between husband and wife, like all relationships of power, was fraught with tension. Husbands and wives negotiated their conflicts in the language of family, often invoking affection and demanding fulfillment of mutual obligations. But the clear asymmetry of obligation within marriage, the surfeit of privilege accruing to the husband in contrast with the numerous obligations of the wife, prompted occasional displays of tart humor. One newspaper story titled "Man's Love," featured a dialogue between two women. When the subject of male love came up, the more cynical of the two laughed out loud and told her naive companion, "When a man finds his house in disorder, and wants somebody to put it in rights, he calls this love; when he is alone too, and things don't go pleasantly, and when he wants somebody to complain to and find fault with, and lay the blame upon, he calls this love. When no one cares for him, and he gets put down in society, and wants to bind himself for life to some being who will flatter him, and admire his very faults, this too he calls love.—Man's love, indeed!"[20]

Husbands and wives invoked affection outside the pages of newspapers as well. Lizzie Scott Neblett of Grimes County was typical of many women who felt perpetually put upon with household and child-rearing duties. In Lizzie's case, her burden was made heavier by her husband's frequent absences from home. At one point, while her husband Will was away, she wrote to complain that his letters were not loving enough. Will defended himself by telling her, "This I think unjust and unkind." He continued, "But admitting that I spoke in it no word of affection or endearment[,] has my conduct in this respect been of the opposite kind as even of such an equivocal nature as to leave the question of my affection to turn upon words of declarations of mine[?]" Her grievance had much to do with her overwhelming household duties; her mode of expression was the language of familial affection.[21]

Despite shouldering a surfeit of the obligations in a marriage, wives found that they could occasionally defy the wishes of their husbands, as long as their defiance lay within the boundaries of wifely prerogative, in areas such as child-rearing. Although most Anglo women along the Brazos did not reject the notion that it was their duty to bear and raise children, they did not necessarily believe they were obligated to bear an endless stream of them. When Lizzie Scott Neblett found herself pregnant with her fourth child, the prospect of going through childbirth brought her to the point of depression. In a letter to her cousin, Jennie Teel, she confided that she had contemplated suicide. Jennie responded in shock and reproved her cousin, validating her complaints but also reminding her of her familial obligations. "It is awful, the pain of child bearing, before and after, but then we can bear it, and ought to be resigned," she wrote, "especially when we have a kind and tender husband, and a good and comfortable home." Jennie's letter seems to have banished all thoughts of suicide in Lizzie, but it did little to ease Lizzie's anxieties about childbirth. Finally, Jennie suggested that Lizzie solve her problem by using birth control. "It is in the power of every woman to prevent children coming, tho' every woman does not know the preventive. . . . 'The Sponge' will prevent conception," she told Lizzie, revealing that she did not deem it a great violation of duty for a wife to limit the number of children she bore.[22]

Although it was the planter class, along with allies such as urban professionals, newspaper editors, and Protestant ministers, who were in the best position to define the concept of family, non-elites such as yeomen and wage hands partook as well. Nonslaveholder households may have looked different from those of the planters, tending to consist solely of nuclear kin, but non-elite Anglo spouses linked affection and obligation in the same manner as their wealthier neighbors. Yeoman couples left no letters revealing their intimate conversations, but we can glimpse something of their relationships through divorce records.

Judicial divorce, established in Texas in 1841, was used almost exclusively by non-elite couples—very few slaveholders risked their social standing by seeking formal separation.[23] Divorce petitions, it must be acknowledged, are not unmediated documents: most were composed by lawyers, so they do not necessarily represent the actual words of the petitioners. Moreover, they necessarily speak only to specific points of law and are calculated to win the sympathy of a jury frequently composed of wealthier, taxpaying neighbors. Nevertheless, through careful analysis, we can learn a great deal from divorce records, because the petitions often deviated from the standard legal format to include what was obviously personal language. Thus, when George Nixon told the court that he had believed he was "uniting himself for life with one of a congenial spirit, whose delight it would be to share with him the cares and comforts, the joys & sorrows of life, one whose friendly smile & [illegible] encouraging voice & counsel would soothe & strengthen his heart when troubles come down . . . & nerve his right arm with power to fight manfully the great battle of life," we can probably assume the sentiment was his. It is in these words and images that we can glimpse the logic of Anglo yeoman marriages. An examination of divorce records suggests very strongly that Anglo yeomen spoke a similar language of familial affection and shared with their wealthier neighbors a common vocabulary of power and hierarchy.[24]

Among southern states, Texas had quite liberal divorce laws, part and parcel of the effort to order marital practices following independence. The law provided for divorce without making any distinction between divorce *a mensa et thora* ("from bed and board," or a separation in which the parties were not allowed to remarry), and divorce *a vinculo matrimonii*, which meant that both parties could remarry.[25] It also gave responsibility for adjudicating divorces to the county's district court. Several other southern states permitted divorce only by act of the state legislature, clearly a much higher hurdle than securing a divorce from a judge. Grounds for divorce included the "natural or incurable impotency of body at the time of entering into the marriage contract," which underlined the reproductive function of the household and the need for family laborers. A husband could obtain a divorce if his wife "shall have voluntarily left his bed and board for the space of three years." Wives, though, could obtain divorces only "where the husband shall have left her for three years with intention of abandonment, or where he shall have abandoned her and lived in adultery with another woman." In Washington County over a period of eighteen years, the district court heard 42 cases of divorce (see table 7). Of these, 25 were brought by women, and 14 of the 25 were for abandonment. Here the law was quite straightforward in granting divorce after a period of three years, and in 11 of these cases the divorce was granted. Of the remaining 3, 2 were dropped by the plaintiff, and 1 was denied on a technicality. Far from presenting a challenge to the region's domestic hierarchy, in granting these divorces the court seems to have endorsed its basic principle, that the head owes his dependents subsistence and protection. As each divorce decree emphasized, the ties that had formerly existed were forever dissolved, and each party was free to marry again.

More revealingly, in a provision rare in other southern states, Texas allowed either partner to sue for divorce based upon "excesses, cruel treatment, or outrages towards the other, . . . [so] as to render their being together insupportable." The clause spoke to the very ideal of the companionate marriage and the notion that families ought to be knit together by affection, and nonslaveholding petitioners took full advantage of it.[26] Significantly too, a surprising number of petitioners charged their spouses with cruelty when suing for divorce on different grounds, to no legal purpose at all. Rachel Crabtree, who already had an airtight abandonment case, charged that husband James "neither loved nor respected her." Livingston Wilkinson, who also had a clear case of abandonment, said of wife Mahala that, "instead of a help-mate & affectionate companion, he found he was wedded to a woman of the most ungovernable passions . . . a perfect fury." And Elijah Pennington, who accused his wife of adultery, added that she "entertained none of those feelings . . . which a wife ought to have for her husband."[27] These cases aside, between 1839 and 1857, 8 women sued for divorce solely on the grounds of cruel treatment. Of these, 4 were successful, 4 dropped the charges, and none were denied. Of course, women married to violent men sought legal redress only at great risk to themselves and possibly their children, so we should not conclude that all abused women took advantage of the law. In addition, we must look critically at the number of suits for cruel treatment that were denied, which was zero.

This suggests that those women who could not meet the legal standard or muster supporting evidence never even bothered to file. Finally, four women dropped their cases for reasons unknown. The possibilities are many: fear of retribution by angry husbands; the desire to avoid a public airing of domestic complaints; use of the legal process as a negotiating tool; and cooling-off and forgiveness.

Still, four women managed to convince juries that their husbands had been so cruel as to warrant dissolution of the marriage. Two of the four alleged that their husbands were drunkards and stressed that they had lost the ability to provide for their dependents. Mary Thompson's husband, John, had beaten her repeatedly, stopped working, spent her separate estate, and left his wife "in poverty."[28] Mary Wells, who married Edward in 1843, secured a divorce nine years later for his "drunkenness, gambling, and debauchery . . . [and] acts of fornication and debauchery with negro women." And like Mary Thompson's husband, he had squandered her separate property.[29] In such cases it is difficult to say exactly which discrete acts the jury believed had rendered married life "unsupportable," in the words of the law. Drunkenness, gambling, and fornication with slave women were certainly frowned upon, but it was the husband's inability to meet his obligations—making married life literally "unsupportable"—that struck at the heart of family ideology and probably swayed the court.

But judges were sympathetic to interpretations of cruelty and did not necessarily require proof that a husband could no longer provide for his wife. In one charge to a jury, Judge J. H. Bell said that cruelty may consist of excesses of treatment, such as cruel language, and both men and women seized upon the idea. Mary Starnes obtained a divorce from Arthur Starnes by invoking norms of family behavior. She told the court that Arthur had slandered her and forced her to leave their home and stay with her parents, treatment she characterized as "worse than that of an ordinary Master to his slave." Because Arthur Starnes failed to appear in court, Mary received a default verdict granting her divorce. Whether a judge and jury would have responded favorably to her assertion that she had been treated like a slave remains academic, but allowing for the dissolution of one purportedly organic relationship because it resembled another contained considerable potential for trouble in a slave society.[30]

One last case suggests that some juries at least understood cruelty in ways that had little to do with the husband's failure to fulfill his material obligations to his dependents. Sophia Lewis charged that her husband, William, was a drunkard who had beaten her and threatened to kill her. Although he admitted he was not "perfect," he denied that he had been cruel. The court granted the divorce, but finding William to be a fit parent, it awarded him custody of the couple's three sons and Sophia their two daughters. Displaying the utmost confidence in William's ability to support a family by granting him custody of his sons, it leaves only the "cruelty" as the reason for the divorce. In judging William's threats and violence to be serious enough to warrant ending the marriage, the court expressed the notion that there were limits to the degree and severity of violence used by husbands toward their wives, and it confirmed the notion that affection was the proper disposition for a marriage.[31]

Imperium in Imperio

If in the early years of colonization the family rights of free migrants were clouded in ambiguity, the rights of the enslaved were even murkier. Had Anglo colonization actually begun under Spanish sovereignty, the legal standing of the slave family would have been clearer. According to *Las Siete Partidas,* the comprehensive body of thirteenth-century civil law that formed the basis for Spain's imperial slave code, and according to Catholic Church doctrine, slaves could legally marry, even against the will of their masters. Not only that: in cases where two spouses had different owners, the law encouraged masters not to separate them, going so far as to prohibit a slaveholder from moving away with a married slave "because they [the couple] would have to live apart." Needless to say, masters seldom felt obliged to follow these rules, and the annals of slavery in the Spanish Empire are replete with examples of Church and state officials failing to enforce the regulations, but Brazos slave families would have enjoyed at least some standing and in some instances might have availed themselves of legal or Church protection had the laws been preserved.[32]

When Mexican independence arrived in 1821 under the auspices of the "Three Guarantees" (one of which was the preservation of Roman Catholicism as the established church), no meaningful measures were taken to extend the existing slave code. Presumably this was the case because the architects of the new Mexican state saw no need to regulate a moribund institution. Slave families, like the institution itself, fell into a juridical limbo that eroded whatever rights existed under Spanish sovereignty. In 1825 the legislature of

Coahuila y Texas passed a colonization law containing an article that decreed, in reference to the introduction of slaves, that "new settlers shall subject themselves to the laws that are now." The wording suggests some continuity with the earlier codes, but officials in Saltillo told Stephen F. Austin that they interpreted it as applying only to earlier prohibitions on the Atlantic slave trade, not to the institution itself, and thus they failed to extend Spanish legal protections to slave families. At various other times in the Mexican era, slave families received indirect recognition, as in Article Thirteen of the 1827 Coahuila y Texas Constitution, which provided for the free birth of any child born to a slave mother six months after ratification, or in a law passed that same year requiring local ayuntamientos (municipal councils) to register all births to slave mothers and to educate all freed children. State officials never enforced these measures, partly because they tolerated slavery as the only practical means of development and partly because of the institution's confused legal situation. After all, how could the state regulate an institution whose legality was in doubt? Ironically, despite the antislavery impulses of most Mexican authorities, the ambiguous official silence on the matter allowed Anglo American practice to become the custom of the country and ultimately helped to extinguish any family rights the enslaved might have enjoyed.[33]

The failure to enforce slave family rights in the early years of Brazos settlement did not mean that there were no slave families. Slaves pressed their demand for family rights ceaselessly and maintained family relationships even when forbidden by their masters. Yet if the primary impetus for family formation emanated from the enslaved, masters perceived advantages to it as well. Tolerating slave families could of course result in an increase in the owner's property. It also opened new possibilities in the perpetual contest of wills between master and slave. Brazoria County's William Russell found out just how serious slaves could be where it concerned their own relationships. In 1837, after hiring a group of eight men from Ashmore Edwards, brother and partner of the slave smuggler Monroe Edwards, he desperately wrote back to Edwards, "I am totally at a loss to know how to manage with them." The problem was that Russell had used trickery to get the men away from their wives, and now the hired slaves had fled the plantation. As Russell summarized the negotiation process for Edwards, "If their wife are taken from them again, unless they are kept in chains they will follow there, and take them with them in the woods, fully determined to live there and not to serve either yourself or me."[34]

There were definite obstacles to the formation of family bonds. The most significant of these concerned the slave population and whether it was large enough and concentrated enough to permit independent family formation. Almost from the start of settlement, there were enough Brazos slaves living on holdings near enough to each other to permit some semblance of family life. As early as 1826, approximately three-quarters of the colony's 443 slaves were clustered in what eventually became two Lower Country counties, Brazoria and Fort Bend, mostly on holdings that fronted the Brazos River. Upper Country slaves were also geographically concentrated, with most living on a single large plantation, Jared Groce's Bernardo, where family groupings were the rule.[35]

Unfortunately, the 1826 census provided no information on sex or age, making it impossible to know whether it was demographically possible for most enslaved residents to find spouses. Anecdotally, the few surviving estate inventories all suggest that the sex ratio was balanced enough for many slaves in the 1820s and 1830s to find mates, even if they could not always form enduring unions. An 1831 inventory of Jared Groce's Bernardo Plantation grouped his 117 slaves into 26 household units, 17 of which were comprised of two parents and children, 8 containing a mother with children, and 1 composed of an adult man living alone. The surplus of adult women is somewhat surprising in light of the supposed reliance on male laborers in the breaking of new land. Groce may have done as many other migrant planters did and purchased women to help grow his labor force, or it may be that he took equal numbers of men and women to Texas, only to lose more of the men to flight. Family formation was certainly easier on a large plantation like Groce's, but it was not unknown on smaller holdings. R. H. Williamson's 1831 "divorce" settlement (his was a bond marriage) listed a 25-year-old woman named Clarissa with two children, as well as a 35–40-year-old man named "Black Tom" and a 55–60-year-old woman named Violet. Another early inventory, that of Henry William Munson, dating to 1833, lists two female-headed families, with six and four children, respectively, and one male-headed family with four children. Together, these inventories suggest that two-parent nuclear families were common enough in the early years, and where they were hampered it was due to a shortage of men, not women.[36]

After independence in 1836, the legal status of the slave family was finally clear: it had none. The constitution written that year for the Texas Republic established continuity with practices in the United States, stipulating that "all persons of color who were slaves for life previous to their emigration and who are now held in bondage shall remain in the like state of servitude," while guaranteeing the right to import slaves from the United States and decreeing that all future slave migrants would be held "by the same tenure by which such slaves were held in the United States," that is to say, as chattels personal without family rights. The only mention of slave family relations in the law after 1836 appeared in an 1858 statute allowing free blacks to choose a master as part of a process of voluntary enslavement. In such cases, the mother, as a free petitioner, had the right to enslave all of her children under age fourteen to the same master she chose for herself, although following enslavement there was nothing to impede separation. Needless to say, few if any free women of color (and there were very few of these on the Brazos) ever enslaved themselves. The law stood largely as a monument to the vanity of a master class steeped in its own racial and proslavery rhetoric.[37]

Whatever demographic obstacles may have hindered family formation in the earliest years of settlement had largely faded by 1860. The main reason for the change was the dramatic increase in the slave population, especially after 1850. The numbers alone would have facilitated partnerships, but the growing size of individual slaveholdings along the Brazos, which averaged seventeen by 1860, would have made it easier to find spouses on the same plantation and made abroad marriages less prevalent. Here, the contrast with Upper South locales is striking. In one county in Northern Virginia, more than half of all slaves lived on holdings smaller than ten; on the Brazos, it was less than one in five. Larger-scale slaveholders in Virginia also tended to divide their laborers over several smaller quarters, a practice rarely seen on the cotton and sugar plantations of the Brazos. The few planters who did divide their labor forces, such as the Mills brothers and Abner Jackson, generally owned enough slaves to stock several large plantations, sometimes with more than one hundred slaves each. Finally, the male-to-female adult sex ratios along the Brazos were relatively even, with approximately 109 men between the ages of fifteen and fifty in 1850 to every 100 women in the same age range, and in 1860 it was 110 men between fifteen and fifty for every 100 women in the same range.[38]

As a result of these changes, it grew easier for slaves to form families. Their families also tended to be nuclear in structure. In the absence of a critical mass of probate records (generally unavailable because large-scale slavery lasted only one generation in Texas), reliable quantification for any part of Texas is virtually impossible. One study of slave families in Louisiana found that 50 percent of all households consisted of nuclear kin with coresident male-female partners. With a cotton-sugar plantation economy similar to Louisiana's, the Brazos probably featured a similar proportion. At the same time, the flexibility seen by all recent studies of the slave family undoubtedly was the outstanding characteristic. One example of slave family adaptability is the high incidence of serial monogamy, the practice of taking several partners over a lifespan but remaining faithful during the period of the relationship. This practice was almost certainly a response to high mortality rates and separation, but it was probably also symptomatic of a different understanding of what family was and how it operated.[39]

Whether black families were animated by the same notion of affection as Anglo families is difficult to say. Because slave owners feared independent religious meetings and generally tried to ensure that white ministers led slaves in worship, contemporary documentation on the interior and domestic spiritual lives of Brazos slaves does not survive. However, slave families were often structured differently from free families and probably operated differently. As the most complete record of any Brazos plantation, the Peach Point Plantation ledger suggests both the strength and the adaptiveness of the slave family. Between 1836 and 1860 there were at least eighteen distinct families in the quarter. Five of these show evidence of serial monogamy. Silvey gave birth to a child in 1845 by Simon, but beginning in 1847 she gave birth to nine children by the same father, Peter. In 1836–39 Clerissey gave birth to three children by a man named Henry from a neighboring plantation. For reasons unknown, the relationship appears to have ended by 1840, because she gave birth to five children all fathered by Simon. (Whether this is the same Simon as the father of Silvey's children is not certain because there were two adult men by that name at Peach Point.) Anney and Dick had one child in 1834; then after marrying London in 1857, she gave birth to five more. Betsey had three by Bill between 1843 and 1847, then two by George in 1851 and 1853, while Julian (Julianne) had one by Bill (again, there was more than one Bill at Peach Point, so it is impossible to say whether he was the same) in 1852 and three by Ned several years later. However, thirteen of eighteen women at Peach Point had children with only one man. All of this suggests that Brazos slaves, as elsewhere, viewed it as acceptable to have children before marriage but also eventually valued the stability of long-term relationships.[40]

This sort of stability required the careful selection of partners. Through such practices as abroad marriage, in which partners with different masters occasionally visited one another, it was possible to create and maintain relationships that might not otherwise have existed. Although abroad marriage undoubtedly strained many relationships, Brazos slaves displayed dedication both to marriage and to emotional compatibility. Joe Lee was permitted only a yearly visit to his wife and children after they were sold to another Brazoria County planter. Ever the provider, Lee brought them a box of groceries on every visit. To be certain, not all abroad marriages held under the stress of constant separation. Sex outside of the marriage—whether engaged in willingly or forcibly—redoubled the stresses placed on slave marriages. Yet, they pursued relationships with the "right" partner, suggesting that marriage was about much more than sex or the production of goods; it was about companionship as well.

Joe Lee's dedication to his wife and children underlines another feature of Brazos slave marriage—ethnicity. In the Lower Country especially, Africans lived and worked alongside creoles, introducing a dimension to marriage patterns that had largely vanished elsewhere in the South. Whenever possible, Africans along the Brazos formed unions with other Africans. The best-documented example of this endogamy was at the Mims Plantation in western Brazoria County. The Mims place had a particularly large African population that dated back to the 1830s, when slave-trader-cum-patriot James Fannin landed a number of captives with his partner, Benjamin "African" Fort Smith. Fannin died a martyr's death at Goliad during the Texas Revolution, but the African community on San Bernard River Plantation remained intact. The land and the slaves were acquired by another Fannin partner, Joseph Mims, in 1837, and after Mims's death in 1845 the estate passed to his wife, Sarah, who died in 1862. With the passing of each proprietor, an inventory was taken, leaving behind three successive probate inventories to document the development of the plantation's slave families, a unique circumstance for the Brazos and perhaps all of Texas. We can supplement the information on the Mims community with data from the federal census. Normally, the slave schedules enumerated slaves by sex and age, not by household. In 1850, however, the census taker for Brazoria County inexplicably grouped slave mothers with their children, making it possible for us to identify them by locating similar groupings in the estate inventories. Finally, it is possible to locate several of the Mims slaves in the 1870 census, the first to enumerate former slaves by name and nativity.[41]

The Mims plantation contained 13 distinct slave households in 1845. At least 7 included at least one African, either the husband, the wife, or both. Six of the 7 families endured until at least 1862, and 5 were still intact for the 1870 census. An eighth family, with an African father, appears to have been formed between 1850 and 1862 and was still in the area, intact, in 1870. The patterns of spousal choice are revealing. None of the African-born women married a creole; all of them formed unions with African men. Of the 8 African-born family men at Mims's, 3 married African women, 2 married women of unknown nativity, and 3 married creole women, though quite significantly, one of the three creole women had African-born parents. Furthermore, this high level of endogamy among African women seems to have been the rule throughout the region. According to the Brazoria County census of 1870, of the thirty-three married or partnered African-born women, twenty-eight appear to have been married to African-born men.[42]

This last detail suggests a strong desire on the part of both men and women to pair with someone of like background. In fact, the high consistency with which Africans sought out African marriage partners suggests not merely that "African" constituted a distinct identity among Brazos slaves, but that African-born men and women of specific linguistic, religious, and cultural backgrounds sought—and in many cases found—marriage partners with similar backgrounds. We have already seen that this was possible, given the documented existence of African speech communities, and given that the structure of the Atlantic trade to Cuba drew heavily on a few areas of Africa, resulting in the transference of specific local cultures and identities to the Brazos. With clusters of people of probable common origin dotting the Lower Country, it is not unreasonable to assume that coherent language and spiritual communities may have formed, with family constituting the anchor of these communities. At the same time, however, the reconstitution and reproduction of African identities along the Brazos had to contend with a serious obstacle: an unbalanced sex ratio among the African-born. Not all African men, even on rather large plantations like Mims's, could find spouses of like background. Although a few found creole partners, the Mims quarters contained six African-born men who seem never to have married, or at least not on the plantation. In the end, it seems reasonable to conclude that Africans did manage to transplant and even reproduce their identities in Lower Country plantation districts like Gulf Prairie, but that a surfeit of men would have kept many from finding suitable marriage partners, leading either to bachelorhood or a gradual merging of African and creole communities.[43]

Slave families functioned as productive households as well. Where it concerned work in plantation crops, such as cotton and sugarcane, even food crops such as corn and sweet potatoes, slaves acted as members of the master's productive unit, with all fruits belonging to the slave owner. But many Brazos slaves, especially on the larger holdings, produced crops for themselves and in that sense constituted semi-independent units of production. Such arrangements, which were not universal but certainly common, were largely the product of insistent negotiation on the part of the enslaved. Many slaveholders tolerated or even encouraged independent production.

Their own domestic world ordered by the language of family, some masters, especially those with evangelical leanings, saw the encouragement of slave marriages as a way to complete the ordering of their plantation quarters. The project was fraught with conflict and contradiction, for as in the old paradox of imperium in imperio, planters implicitly understood that any division of their own household sovereignty would profoundly alter the nature of the slave system. Those slave owners most devoted to the ideal fretted with each intervention in the domestic affairs of their slaves, but intervene they did. When news reached James P. Caldwell that some male slaves on his Gulf Prairie Plantation had abandoned their wives, he told his stepson, who was overseeing the place, "Don't permit Henry or Davy to seperate from their wives & if they misuse them flogg them for every offence."[44]

Individual arrangements ranged from constrictive to quite lenient, with many masters erring on the side of leniency in a bid to discourage flight to Mexico. James F. Perry's Peach Point on Gulf Prairie fell into the latter category. Organizing what he termed his "black family" by male-headed households, he allowed his slaves to earn monetary credits by producing cotton during off-hours and on Sundays. He then marketed the cotton through his own factor in New Orleans, acting in effect as his slaves' local factor. Interestingly, Perry paid ten cents per pound for his slaves' independently produced cotton even when the price at New Orleans was barely half that, further evidence that this system was shaped primarily by the moral-economic demands of the enslaved. Most significantly, Perry dealt only with male "heads" of household, setting up an account for each in his plantation ledger. With each male slave accumulating credits for the extra cotton, Perry allowed them to purchase goods, many of which were clearly intended for their wives and children—shoes, clothing, even a dress pattern from New Orleans. When the plantation converted to sugarcane around 1850, Perry marked off a one-acre plot for each adult male and allowed the men to till the plots for themselves. Similar arrangements were not unusual on sugar plantations. Evidence from Louisiana suggests that sugar planters allowed more independent production than cotton planters, largely as a way to dissuade slaves from spoiling fragile cane. Cotton planters were generally less permissive, while still allowing slaves to produce their own crops. Thomas Blackshear of Grimes County allowed his slaves to produce "coal" (cabbage), which he bought from them, allowing the growers to accumulate monetary credit in his ledger. There were almost as many variations as there were plantations.[45]

In each case, small-scale production was structured around a male "head" of household. It was his name that appeared in the ledger, and through him ran the connection to the larger world of the market. In short, judging from plantation records, the household dynamics of the enslaved appear to mirror those of free Anglo families. But whether they did so in fact or whether they simply did in the planters' ledgers is difficult to say. Ultimately, given the ability of a planter to terminate, alter, or interfere with independent production, it is clear that slave men were not the domestic sovereigns that free men were. Even without the master's interference, it is still not clear that male slaves ruled their households as patriarchies with a patriarchy. On the one hand, there is no evidence to suggest that slave men objected to functioning as heads of household. Moreover, male slaves seem to have embraced the chance to act as providers for their wives and children—recall Joe Lee's annual gift to his abroad wife and children. Slave narratives reveal many more instances of this. On the other hand, it is clear that slaves often departed from the patriarchal script. At Peach Point, adult men frequently helped other adult men accumulate credits. Not only that, but the men enjoyed no monopoly on women's labor—there are examples of adult women working for two or more men. So while it is reasonable to conclude that slave households may have possessed a latent tendency toward male domination, slaves did not always find it practical to organize themselves that way and remained open to alternatives.

The differences between free and enslaved households along the Brazos were further amplified by the existence of African families on many of the larger plantations of the Lower Country. Those born and raised in Africa would have held very different understandings of how households worked. One recent work has suggested, for example, that Yoruba-speakers in Oyo, the likely source of nearly half of the Africans along the Brazos, did not conceive of a world divided between "men" and "women;" that is, they acknowledged no permanent masculine-feminine dichotomy. A person moved back and forth across the boundaries of gender according to context—in spiritual life, for example—rather than according to one's body type. One need not go that far to realize that African women occupied household spaces that were quite different from those in the U.S. South. One of the more significant differences, at least where it concerns production, is in the construction of agriculture as a feminine, rather than a masculine, pursuit. Yoruba women—indeed, women throughout West Africa—were often the principal tillers of the soil. Moreover, many of the crops they cultivated back home were grown along the Brazos as well, cotton, corn, and tobacco among them. The primary exception, and a crop that carried a distinctly masculine identification, was the yam. With this responsibility for agriculture came the right to control the marketing of the crop, and throughout the continent, a great deal of market activity fell to women. Women not only took produce to markets and sold it; they frequently functioned as market regulators, determining fair prices and practices and settling disputes. Africans carried their marketing customs to most of the New World plantation colonies, although the sparse settlement and the absence of significant urban areas precluded their development along the Brazos.[46]

Lacking sources, it is difficult to know what the Africans along the Brazos thought of the new forms of household production and organization they encountered. Whether they discarded old attitudes out of practicality or whether they organized independent production in a manner congruent with Old World experience is difficult to say. One possibility, though hard to document, may be the practice of task-swapping evident in the records of Peach Point, home to at least two Africans. Many West African societies viewed labor as a communal endeavor to be undertaken not by individual households but by kin groups. At Peach Point, a number of slaves performed "overwork" during cotton-picking season and had their picking totals reassigned to other members of the slave community. (For more detail, see chapter 3). In addition, some scholars have suggested that many captives from Yoruba land, wracked as it was by war during the first decades of the nineteenth century, had extensive experience with shattered family life and may have been particularly experienced at creating ties of fictive kinship. This experience would have served them well in the New World.[47] Gendered notions of work may also have informed African households. John Allevouw and Bill "Affrican" were the two Africans who appeared in the Peach Point ledger. Both produced extra cotton and both received one-acre plots to till as their own. Judging from the ledger, there is virtually no difference between John and Bill and the creole heads of family at Peach Point. In it we find no suggestion that they may have found growing cotton and corn to be "women's work," or that they organized production in a manner different from the way their fellow slaves or, for that matter, Anglo yeomen did. Yet, it strains credulity to believe that they could have forgotten all previous attitudes. If we have learned one thing from studies of the African diaspora, it is that the enslaved did not abandon their values or forget the things they knew prior to arrival in the New World. It is entirely plausible, if empirically unprovable, that household production for Africans along the Brazos was organized in a manner more in keeping with the world they left than with the world in which they currently found themselves. The families and households of the enslaved, while resembling those of the master class in certain respects, likely did not, for reasons of legal status, practicability, and cultural differences, mirror them precisely. Slave families and households were characteristically "in-between," along with so many other borderland institutions.[48]

Despots

Different understandings of the relationship between husband and wife helped mark the boundary between the German uplands and the Anglo-dominated bottomlands. The differences were subtle but still attracted the occasional comment. Passing through in 1849, German traveler William Steinert saw Anglo marriages as profoundly influenced by slavery. As he saw it, slave-owning women grew lazy with slaves to do their work for them; nonslaveholding women, anxious about their precarious status as "ladies," came to despise all household labor as well, and lobbied their husbands to acquire slaves. Anglo yeoman women, Steinert noted with disapproval, "sit in their rocking chairs, smoke their little pipes, rock themselves, or at most embroider a bedspread." As a result of the women's indolence, Anglo houses were rather spare and slovenly. "If the women can afford it," he noted, "they live in great style . . . but the people [along the Brazos] do not believe so much in an abundance of washable clothes, [and] tablecloths."[49]

Aesthetics aside, Anglo farmwives worked a great deal more than Steinert gave them credit for. But his fixation on women's work was entirely typical for Germans of his era. Back in Europe, the composition of the household, as well as the meanings and reciprocities among members, had been changing rapidly. In the area surrounding Osnabrück in the Kingdom of Hannover, the practice of impartible inheritance, combined with an increase in population, resulted in large numbers of landless couples. As we have seen, the proportion of landless peasants forced to live as *Heuerlinge*, or tenants with labor obligations to the landlord, soared from about half of all households in 1700 to more than two-thirds for most of the nineteenth century. Since most land-owning peasants did not provide separate dwellings for their tenants, it was not unusual for two couples with children to share a single cottage. Heuerlinge also had fewer children than proprietors did and generally sent their adolescent offspring to work as servants for landowning neighbors. Many households produced linen to supplement agricultural income, growing the flax in summer and spinning it during the winter. Much of this work was done by women, an arrangement that prompted alterations in the ideals of marriage and family.[50]

In regions of partible inheritance, those changes played out somewhat differently, though with comparable significance for household and family. In the southwestern town of Neckarhausen, David Warren Sabean found, a population increase over the course of the eighteenth and early nineteenth centuries prompted intensified agricultural practices, which in turn altered earlier understandings of household labor and obligations. During the years of the greatest stress (which were roughly 1800–1840 in this particular case, although the pattern was repeated in other areas at different times), women forced a renegotiation of family reciprocities based upon "mutual dependence," moving toward a more contractual vision of marriage, rather than a "hierarchical structure with women dependent on men." As agriculture intensified and as households increasingly supplemented farm income with home-produced linens, women found themselves thrust further into the market, the value of their labor enhanced in relation to their husbands'. Family discourse centered more on whether spouses had lived up to their obligations than on whether dependents had obeyed the head.[51]

With family and household relations in flux in Germany, those who migrated to Texas entered a world with still different dynamics of property, labor, and markets. The effects are visible in the evolution of household structure in Cat Spring (see table 8). The contrast with the slaveholders of Gulf Prairie and Chappell Hill (as well as with many German regions) is striking. From 1850 to 1860, both Chappell Hill and Gulf Prairie, neither of which had a large German population, saw a decrease in the proportion of nuclear households, which was closely tied to the increase in the number of slave-owning households. Cat Spring, with its large German population, moved in the opposite direction—the percentage of nuclear households increased dramatically as immigrant families acquired land and established small farms, jumping from 49 percent in 1850 to 76 percent in 1860 (see table 8).

The increase in the proportion of nuclear households in Cat Spring reflects a number of factors, encompassing both the demands of small farm production and problems associated with transatlantic migration. Particularly striking is the number of households in 1850 that contained people outside the nuclear family. Members of the household head's extended family accounted for some (6%) of these, and we can probably assume that many of those counted as "ambiguous" fall into the same category but are untraceable because of different surnames. Anecdotally it is possible to find several examples of siblings migrating together, as the brothers Charles, Marcus, and Friedrich Amsler did in 1834. Migrants, once they became established, sometimes sent for their elderly parents. The creation of households containing extended family was a common immigrant strategy but usually represented a temporary arrangement. As the Cat Spring immigrants established farms, nonfamily and extended family members moved out, increasing the proportion of nuclear households. Unlike the Chappell Hill and Gulf Prairie households, very few households contained nonfamily members—only 2.6 percent for certain, although some of the "ambiguous" households probably contained some as well.

The cotton economy of the Brazos forced German households to reverse some of the adaptations that had begun on the other side of the Atlantic. Women in some German regions, for example, had cultivated flax and spun it into linen for the market, providing many with a source of income, which in some cases led to marital conflict over control of family finances. These opportunities vanished with the move to the Brazos. Cotton culture, which commanded the labor of entire households, pressed German women toward greater dependence on their husbands. Gone was the market for home-produced textiles; gone were significant markets for truck garden vegetables. And in contrast to the increasing number of Anglo women who owned slaves (a majority by 1860), most Germans had none, although it was still possible to secure labor by boarding newly arrived adolescents. In Texas, husband, wife, and children performed most of the labor on a crop that was marketed and controlled by the husband. Thus, with the move to Texas, German women likely experienced a loss of power within the household.

These realities were reflected in household relations. According to William Trenckmann, who lived near Millheim in Austin County (and whose family apparently employed domestic servants from time to time), the sexual division of labor varied somewhat according to the needs of the season. In spite of this variation, a clear understanding of "men's" and "women's" work prevailed. In general, according to Trenckmann, his father worked in the fields and his mother cooked and sewed the family's clothing. When he was old enough, William was assigned masculine tasks while his sister was given feminine ones. "I had never really known hard work," he wrote with more than a hint of irony: "Hunting eggs, chopping wood and bringing it in, the delightful hauling of barrels of water with horse and sled, driving in the calves early in the morning and in the evening the cows when the grass still tasted too good for them to come home, roping off the calves while my sister milked, feeding the hogs, laying cornstalks in the furrows before the plow, leading the oxen before the iron plow for deep plowing—that was about all I had ever done."[52]

But the boundary between masculine and feminine tasks on German farms could be breached from time to time, as Trenckmann's memoir demonstrates. As a young boy he helped in the kitchen "when there seemed to be an opportunity for licking the pans," and when his mother became ill and could no longer work, William joined his sisters in their chores. "I tried myself at washing, ironing, sewing on buttons and darning when no female help was to be had or when in the case of hired help too many linen napkins, table cloths, and too much good porcelain began to vanish," he wrote. In other words, when his family needed him to do women's work, he did, although a little redemptive humor was required to set things straight.[53] In addition, German women and girls almost certainly worked in the fields at certain times of the year, picking cotton and harvesting corn in the late summer and fall, tasks that Charles Nagel of Millheim described as "labor for the old and young."[54]

The discourse of love and affection that saturated Anglo household relations occupied a less prominent position in the German communities. Among Anglos, the growth of slavery, the withdrawal of women and girls from production for the market, and evangelical Protestantism had the effect of placing familial affection at the center of domestic life. The German experience, both in Europe and in Texas, was quite different. Although love and companionate marriage were fodder for popular German literature, at least one historian has suggested that their appeal was limited to a more educated bourgeoisie and that they were not common elsewhere. Analysis of marriage patterns in several regions suggests that concerns for property and social mobility, not romantic love, tended to govern who married whom.[55] In addition, Anglo American understandings of family had been molded in the crucible of republican revolution, with women emerging as indispensable inculcators of civic virtue.[56] Of course, to say that Anglo families emphasized affection to a greater degree than German families is emphatically not to say that Germans did not love each other. Family members simply couched their demands in different terms and negotiated with one another according to a different understanding of power and reciprocity. Charles Nagel described it this way in his memoir: "Our home was impressed with the sense of duty—father almost austere, mother devoted to make it sweet and gentle."[57]

We can glimpse the difference in scattered statements, such as the closing of a letter written by a young man to his father back in Oppendorf, thanking him for his support and sacrifice. "Dear Father," wrote Johann Friedrich Winkelmann, "now I must turn to you. I give you the heartiest filial [*kindlichen*] thanks for all the trouble that you have gone through for me, which is now worth more to me than gold or silver. Yes, once again I would like to thank you heartily." And he signed his letter, "your loyal son."[58] In a similar vein, William Trenckmann recalled that his father, like many German fathers he knew, "was never able to show his love for me by means of a single caress or expression of affection, and yet he loved me dearly and made sacrifices for me which were far greater than I suspected at the time he made them."[59] Both Winkelmann and Trenckmann demonstrate a vision of the family (specifically father-child relations) that emphasized loyalty, obedience, and gratefulness toward their parents rather than the fawning affection idealized among Anglos, north and south.

Patterns of divorce also suggest differences in German understandings of family. German divorces in Austin County differed from Anglo divorces in Washington County in several ways. Most apparent is the lower rate of abandonment. Abandonment cases made up half of all divorce cases among Washington County Anglos but slightly more than one-third of those among Austin County Germans, and none of the German cases involved wives leaving husbands. There are many possible explanations for the difference. Because many German couples had married before emigration, it is possible that those women who came to the Brazos were generally satisfied in their marriages and would not have made the journey if there had been any doubts. Mathilde Koy, whose husband Franz had preceded her to Cat Spring, demanded reassurance of his devotion before leaving home. As important, deserting a spouse once in Texas may well have brought alienation from the German community, a vital resource for immigrants, stranding them in a foreign land with diminished resources. Such would have been especially true for women, a notion supported by the fact that the only three German female plaintiffs had married in Texas, not Europe.[60]

The most common complaint among Germans was cruelty, which was cited in four cases, and the language used to describe it is sometimes quite telling. In the aggregate, the cases suggest that German couples emphasized obligation rather than affection in their negotiations. Barbara Meury, a Swiss who lived near Cat Spring and sued her husband Samuel for divorce, is a good example. According to Barbara, Samuel had engaged in a variety of abusive practices, especially when drunk, ranging from throwing a coffee cup at her to threatening to kill her with an axe. In the spring of 1858, the Meurys planted a crop together, but soon afterward Samuel left home to work at Friedrich Engelking's and Marcus Amsler's. He apparently returned sometime later and drove Barbara from the house with death threats. Barbara then attempted to return and claim what she viewed as her share of the corn and potatoes they had raised together, only to be chased away again. Samuel then ruined the garden by tearing up all the vegetables. In addition to the struggle over who owned the crop, Barbara Meury charged that Samuel never did "show any parental love" toward their children, which at least in her eyes was part of a generally "despotic" course of behavior.[61]

Samuel Meury saw things differently and contested his spouse's version of events in court. In response to Barbara's petition, Samuel charged that his wife had been unfaithful to him and that their youngest child had been fathered by Henry Seeburger, the husband of Barbara's sister Sophia. He also asserted that his wife's true goal was to claim a five-hundred-dollar inheritance of his and to this end was trying to have him declared non compos mentis. (How a divorce would have figured into this plan was left unclear.) And finally, he denied that his wife had contributed to the planting of the 1858 crop at all.

A recurrent theme in both of the Meurys' accounts is the issue of who contributed to the making of the crop. In Barbara's telling, the crop was the joint product of husband and wife. Samuel, of course, denied that Barbara had contributed at all, refused to allow her a share, and even destroyed the garden (probably a kitchen garden separate from the corn and potato fields). Here, then, lay what was probably a central issue for German couples: who contributed what to the household and what privileges that contribution imparted. For men, who back home had seen their power in marriage diminish with the rise of stall feeding and the intensification of agriculture, the situation in Texas offered the possibility of regaining lost ground. In Samuel Meury's eyes, his wife's claim on the couple's crop was an indication that she was out of control, leading to the conclusion that she was carrying on an affair with her brother-in-law. (Numerous deponents in the case denied the possibility.) For Barbara, her husband's refusal to acknowledge her part in making the crop (and possibly by extension her right to decide how to dispose of it), as well as his violence and lack of affection toward his family, represented an illegitimate, "despotic" assertion of power. Members of Anglo families, in contrast, negotiated with one another according to a discourse of love and affection. In German families obligation lacked the same veneer of love, dwelling instead on duty alone, a notion clearly being contested by Barbara Meury.

This point is further illustrated by the three remaining suits brought by German women on the grounds of cruelty. None mentioned a lack of affection in their petitions or complained of any other sort of mental cruelty, focusing instead on cruelty of other sorts. In 1856, Wilhelmina Buntzel asked for a divorce from her husband Frederic because he had slandered her in public, accusing her of adultery in front of customers at Ferdinand Engelking's store in Cat Spring. Wilhelmina's complaint, interestingly enough, was based not on the notion that Frederic had done emotional damage to her, which would have indicated a discourse of affection, but that he had injured her reputation in the community. For his part, Frederic denied that Wilhelmina's charges warranted divorce, but an Anglo judge and jury, perhaps reading a betrayal of affection into the case, dissolved the Buntzels' marriage.[62]

Two other petitioners complained of physical cruelty, which in both cases involved beatings with a whip. While Louisa Maria Luhn was in confinement after giving birth to a daughter, her husband Frederick beat on the walls from an adjoining room and distressed her "so as to endanger her life." He also refused to see the child, referred to Louisa as "the vilest of her sex," and, apparently with his children from a previous marriage either watching or participating, whipped her. When it was clear Louisa was suing for divorce, Frederick apparently attempted to sell the couple's property rather than permit a division. The divorce was granted. In another case, Frederike Gaul filed a complaint against her husband Frederick, charging that he had "beat and bruised" her with a whip, but for unknown reasons, the Anglo jury found her allegations untrue and denied the petition. When she tried again two years later, the result was the same.[63]

German men also appear to have been different from Anglo men in their understanding of marriage. Although the four cases brought by German men are an admittedly small sample, the grounds cited for divorce are nevertheless revealing. Most Anglo men who sought divorce did so for abandonment. None of the German men did so. Two of the four Germans sued for adultery, exactly the same number as among Anglo men, although proportionately much higher. If, as the scholarship on the cult of southern honor suggests, Anglo men shunned formal justice and favored direct, personal justice, German men appear not to have had the same attachment to the concept. When Joseph Habermacher told the court that his wife, Elisabeth, had had sex with "a number" of men and had given him a venereal disease, he was confessing to a lack of mastery over his own household that few Anglo men would probably have admitted. Naturally, the court granted the divorce, which was in any event uncontested by Elisabeth.[64]

One of the male petitioners sued his wife on the grounds of cruelty, matching the total among Anglo men in a much larger population. As with the German women who sought divorce on the grounds of cruelty, Christian Streapa had nothing to say about his wife's lack of affection. Sophia Streapa's cruelty consisted of repeated public accusations of incest with a daughter from an earlier marriage. According to Christian, Sophia even admitted that she hoped to destroy his reputation and ruin his livelihood. When Christian Streapa sat up one night with a sick relative, his wife met him on the road home the next morning and "abused" him all the way home. For the most part, Christian

Streapa's understanding of cruelty centered on the public nature of his wife's accusations, as well as her stated goal of destroying his reputation and livelihood, not on the loss of her affection. The court agreed with his interpretation and granted the divorce.[65]

German and Anglo marriages were similar in many ways, but notions of marital reciprocity were not among them. Studying divorce in nearby Colorado County, one historian noted that Germans and Anglos filed for divorce at roughly similar rates.[66] Yet if we read the petitions as cultural, as well as legal, documents, some differences emerge. Most notably, affection seems to have had little expediency in German household politics, not surprising perhaps given the less-than-universal acceptance of romantic love as a reason for marriage in much of the German-speaking world. David Warren Sabean has suggested that a discourse of the house, referring among other things to standards of orderliness and cleanliness, rather than affection, became the focal point of marital reciprocity, at least in one region of southwestern Germany.[67] German couples most surely loved each other; but affection lacked the power to oblige in other contexts that it had in Anglo family life.

The settlement policies of Mexico yielded a complex geography of household and family arrangements. This diversity affected Brazos society in two important ways. First, given the different legal and economic influences, households in different Brazos neighborhoods were structured, and therefore functioned, differently. Anglos, Africans, creoles, and Germans all had different experiences of marriage and family. The various understandings of family, however, did not preclude the possibility of learning another familial language, as the enslaved did out of necessity. Second, family was just one of many things that marked the internal boundaries of the Brazos. The denial of official family rights to the enslaved was closely tied to the construction of racial difference. In white eyes, Negroes were promiscuous, felt no familial urges, and were clearly different from Anglos, who knew the exquisite sensation of affection for one's spouse and children. The boundary between German and Anglo families was less apparent. For a few observers, such as William Steinert, distinct family practices really did define German ethnicity: the boundary between Anglo and German settlements was marked by austere fathers and hardworking mothers. But Steinert's perceptiveness was unusual. While we have seen that Anglos and Germans had different understandings of family, very few members of either community seem to have noticed or commented. The boundary between German and Anglo identities was not marked by the language of family, but by other differences, which we will examine shortly.

Chapter 3

MASTERS AND SLAVES

Brazos slaves and slave owners inhabited very different worlds, but the boundaries that marked these worlds rarely received mention in travelers' descriptions or the correspondence of the master class. Travelers tended to view the region as an undifferentiated, integral whole, which might be characterized simply as "the Brazos plantations," with slaves finding their way into the public transcript not as human agents but as proof of a larger social corruption. On an 1828 visit, José María Sánchez marked his passage into the Brazos bottoms by commenting on the "irregular and desultory manner" in which the town of San Felipe was laid out, the "wretched little stores" that dispensed whiskey and lard, and the "drunk American" with his "three intoxicated negroes" who sang continuously and discordantly as they ferried him across the river. Sánchez, like most elite Mexicans of his generation, found slavery distasteful, and the decadence and slovenliness he saw along the Brazos only confirmed his feelings. Nearly thirty years later, Frederick Law Olmsted made a similar observation after supping with the owner of forty slaves. The man's son, Olmsted recalled, was "already, at eight years old, a swearing, tobacco-chewing young bully and ruffian," who could be heard "whipping his puppy behind the house, and swearing between the blows."[1]

Both Sánchez and Olmsted hoped to highlight the corrupting influence of slavery on Brazos society as a whole. But in elaborating a more or less holistic vision of master-slave relations, even from a position hostile to slavery, they missed the essentially conflictual nature of the relationship. The master-slave relationship along the Brazos was not an organic one; rather, it was a perpetually contested one, punctuated by clashes, negotiations, and tactical advances and retreats on each side. The space where the demands of the slave owners met the acquiescence of the people they owned constituted a middle ground, or borderland, bridging the two worlds.

"Mexico in His Head"

Masters and slaves negotiated with one another in this space through the language of family relations, what some historians have termed "paternalism." In characterizing paternalism as a language and not ideology, I hope to avoid the implication that slaves internalized the view of the master class.[2] Instead, I suggest that the image of family lent structure and form to master-slave negotiations, without implying that either party necessarily subscribed to the notion that slaves were to their masters as children were to their fathers.[3] In the seaboard South, a paternalistic, familial metaphor had emerged as the touchstone of master-slave negotiation by the early nineteenth century, the result of several historical developments: the cultural nearing of masters and slaves during the latter half of the eighteenth century; the emergence of a coherent, science-based vision of race, which served both as a theory of history and an organizing principal for all humanity; the rise of evangelical Protestantism, with its emphasis on love and humility; the ascendance of new discourses of emotion in domestic life, which slaveholders extended to their slaves; an intimacy born out of a resident (as opposed to absentee) master class; and the need to justify bondage in the face of growing criticism. The slave owners and slaves who began settling the Brazos in the 1820s were already quite fluent in the language of seaboard paternalism upon arriving in Texas. Once there, masters and slaves, through their lived experiences, created a new offshoot that we might term "borderland paternalism."[4]

As a language, paternalism might best be compared to a pidgin, a specialized language that often emerges in places where people of different backgrounds meet to trade—in other words, a border language. Although pidgins may have extensive vocabularies, they tend to be used for a narrow set of topics, such as trade, where there is a pressing need for communication. Those aspects of life separate from trade never develop vocabulary, making pidgins insufficient for expressing anything more than the narrowest spectrum of human experience. For this reason, pidgins do not have native speakers; they are used only in negotiation and rarely spoken out of context. There are examples of pidgins' evolving into full-fledged languages of their own, termed *creoles* by linguists, but paternalism underwent no such elaboration. To the end, it remained stunted, incapable of conveying any deeper truths about human relations.

Paternalism was composed of an array of words, metaphors, narratives, and parables, all designed to reconcile the contradictions of the master-slave relationship, its speakers perpetually playing off such tropes as "affection," "gratitude," "my family, black and white," "chastisement," and "correction" (the latter two invoked almost exclusively by slave owners). In a true pidgin, the native tongues of all parties contribute words, but English furnished almost all of the vocabulary for paternalism, bearing testimony to the asymmetry of power inherent in the master-slave relationship. That same imbalance of power allowed planters to force slaves to converse in a language not of their own choosing.

Paternalism demanded that slaves flatter their masters. The Perry slaves revealed themselves to be artists with fulsome praise when they asked Eliza Perry to convey various messages and greetings to Stephen, her older brother and the designated heir to the family estate, who was away at school in Ohio. It seems all of the "black ones" were inquiring about him and wanted to know when he would be back. Mary wanted Eliza to tell her eventual master, "God Almighty save him no let him be no sick no tired no nuffing and tell him howdy, me no forget him me talk about him all the time." Peter wanted Stephen to know that he was taking good care of his dog, while Sam told Eliza that "Master Stephen is the most adventuresome person he ever saw, that you just stand under the tree where a panther is, and shoot and never be scared he says he would not be so venturous for anything."[5] They also learned to make demands and manipulate symbols in the language of domestic and familial obligation.

By the time slavery began to spread to Texas in the 1820s and 1830s, seaboard paternalism had formed a number of articulations with other ideas, links that only grew stronger through the antebellum period. There was, for example, a consonance with the political ideology of the early republic, in which slaves were imagined as dependent members of citizens' households, incapable of acting as public-minded, self-sacrificing members of society. Evangelical Protestantism, which in the eighteenth century actually condemned bondage, furnished another connection. By the time Austin began settling his colony, evangelical leaders had made their peace with slavery, and from there it was only a short step to imagining slavery as an unadulterated good because it brought members of a heathen race into the Christian fold. And, as we have already seen, the rise of a new model of family affection, specifically pertaining to fatherhood, offered a parallel that few slave owners could resist.

As masters and slaves migrated into the Texas borderlands, the language of master-slave negotiation acquired new valences that set it apart from its seaboard origins. For masters in particular, the notion of a wilderness, with all its associations, enriched the paternalist vocabulary. For the Anglos who came to the Brazos in the 1820s and 1830s, a sense of the wilderness and all it encoded saturated their every thought and deed. For most if not all of these settlers, wilderness implied first and foremost a dividing line between civilization and savagery. Letters and travel narratives from the period abound with references, both direct and oblique, to this boundary. Mary Austin Holley, a cousin of Austin's who visited the Brazos in the early 1830s and published the first extensive description of his colony, projected the dichotomy onto, of all things, a beehive. The bee, she maintained, "is never found in a wild country, but always precedes civilization, forming a kind of advance guard between the white man and the savage."[6] Needless to say, the presence of bees in Texas confirmed in Holley's mind that things were progressing as they should. Embedded in this simple binary of civilization and savagery was an almost infinite array of analogs, combinations of which lent shape to the elemental categories of self and other: Christian-heathen, masculine-feminine, white-black, lawful-lawless, orderly-chaotic, and chaste-profligate. Far from anticipating a Turnerian escape from repressive institutions, most Anglos actually hoped to transplant the hierarchies that made up their civilization. Put differently, civilization implied an orderliness based upon maintaining the institutions of seaboard society. Who could imagine civilization without the racial and gender subordination it implied?[7]

Among southerners, who were beginning to perfect a proslavery argument just as the settlement of Texas began, civilization also implied racial slavery. Generations of southerners had in fact never known "civilization" without slavery, and in many intellectual circles the two were inseparable. Although the most systematic arguments linking slavery and civilization date from the 1830s, they merely represented a more articulate statement of sentiment that had been cohering for at least two decades.[8] Contemporary statements reflect an aphoristic common sense on the topic. "*Servitude* is the condition of civilization," William Harper of South Carolina declared boldly in 1838; and stronger still: "Slavery is a principal cause of civilization." Civilization, according to Harper, required a division of labor, a separation of the "intellectual" from the "laborious." Without slavery, "vast tracts of the fertile portion of the earth's surface" would remain outside the "pale" of orderly society. Without slavery, progress was impossible, and man would be doomed to "roam the earth as a biped brute." For Harper, small-scale yeoman agriculture was but a stage between savagery and the higher civilization presumably enjoyed in seaboard plantation areas. Small, subsistence-oriented farming was clearly superior to "savage life," but inhabitants of the "more thinly peopled portions of our own country" lived in "the rudest agricultural state." Another well-known southern intellectual, Thomas R. Dew of Virginia, tied civilization to slavery implicitly as he invoked biblical support for bondage. "When [the Israelites] conquered the land of Canaan," he argued in a famous defense of slavery, "They made one whole tribe the 'hewers of wood and drawers of water.'"[9]

Anglos who saw Canaan or Eden in Texas—and there were many—could presumably make similar linkages. Among the more prominent of these was Stephen F. Austin. A native of Virginia and son of a Connecticut Yankee, Austin was no proslavery ideologue in the mold of Harper or Dew. His views were closer to those of the revolutionary generation, which saw slavery as a political evil rather than a positive good.[10] Austin's main priority, however, was the development of his colony, and this led him increasingly to embrace slavery as a means of "civilizing" his piece of the Texas wilderness. Under his first contract, Austin sought and received support for extra grants of land for those who brought in large numbers of laborers, which in practice meant slaves. Later, fretting about the large numbers of landless migrants, "ignorant, self willed, 'mobish' mountaineers and frontiersmen who 'hold to lynch law,'" and other rowdy civilizers, he hoped to attract "Southern Gentlemen" to offset their influence.[11] Slavery, he believed, might also have a salutary effect on Texas Indians by promoting agriculture, which in turn would provide them with a model of civilized life. "If humanity is disgusted by this idea," he wrote in an effort to convince skeptical Tejano and Mexican officials, "I respond that it serves the cause of humanity, on the one hand, to rescue this country from the barbaric inhabitants, and to hasten the epoch of civilizations for these unlucky sons of the American soil [i.e., Indians] bringing them closer, in degrees, to dedicating themselves to agriculture." Not only that, Austin argued, but bringing slavery to Texas actually benefited the slaves themselves, because "the lot of these would be improved by moving to this new country, where there is abundant living, and the lot of their descendants would be improved by wise legislation." Eventually, in 1833, after wavering on the question for some time, Austin concluded that "Texas *must* be a slave country. Circumstances and unavoidable necessity compels [*sic*] it."[12]

Many a planter flattered himself with the notion that slaves would participate as willing junior partners in bringing civilization to the Brazos wilderness, then dutifully take their places as enslaved laborers. How else can we explain the confidence with which Jared Groce, a native Virginian, late of Alabama, led a column of more than ninety slaves into Mexican Texas in early 1822 and established Bernardo, Texas's first plantation, on the bluffs above the Brazos in present-day Waller County? It apparently never occurred to Groce that leading that many slaves into a region where the master-slave relationship enjoyed no legal support constituted a risk to life, limb, or property. And indeed, Bernardo proved to be a profitable plantation, yielding a handsome cotton crop within two years, which Groce's own enslaved teamsters carted into the interior of Mexico for marketing. In 1824, when Austin waged a war of expulsion against the Karankawa people of the Lower Country, Groce armed thirty of his male slaves and rode with them as a self-contained cavalry troop. Up to his death in 1839, Groce acted the wilderness patriarch, directing his bondsmen to build an impressive plantation house surrounded by lesser buildings, including a hospital, bachelor's quarters, and a slave quarter.

As they did with other varieties of paternalism, slaves reinterpreted the paternalism of the Texas borderlands in ways that did not always comport with their masters' vision. Jared Groce, for example, may have believed that his slaves carved out his plantation and rode against the Karankawas because they accepted the role of junior partners in civilizing a wilderness, but doing so would have required them to accept the racist, triumphalist vision of their masters, something we can be reasonably certain they did not accept. We do know, for example, that some slaves forsook the comforts of "civilization" for life among Texas Indians. Slaves in the southeastern United States had done the same since the colonial era, and groups like the Creeks and the Seminoles were famous for assimilating large numbers of runaways. However, it seems that Native American groups in Texas attracted fewer runaways because, unlike their counterparts in the Southeast, most Texas Indians were semisedentary or nomadic. Fugitives undoubtedly found the sedentary agricultural practices of the Southeast more attractive and the people more welcoming. In fact, the evidence suggests that most Texas slaves saw Indians as a threat, not as potential allies. Speaking of Grimes County during the 1850s, where native groups had been largely driven out, Lizzie Atkins told a WPA interviewer, "We had a terrible time in Texas in them days, for the Indians would kill our stock and steal everything we had."[13] Statements like Atkins's help explain why Groce's men would willingly carve out a plantation and join him in expelling the Karankawas: African Americans on the Brazos were in fact building a civilization that they claimed at least partially as their own (despite the fact of enslavement), one that did not imply the same white supremacy as their owners' version.

In addition to its articulation with the notion of a wilderness, borderland paternalism had a second distinguishing component: the border with Mexico. It took longer for the border to work its way into master-slave relations than it did for the concept of a wilderness. The wilderness was an idea familiar to generations of white and black southerners; borders, at least pertaining to slavery, were relatively new. To be certain, slaves had long taken advantage of imperial boundaries to secure freedom, especially in times of political tension. Shortly after the acquisition of Louisiana by the United States, a number of Louisiana slaves fled into Spanish Texas, where they received aid and comfort from government officials.[14] But as late as the 1770s, slavery was legal throughout the hemisphere, a virtually unbounded institution, and outside of the northern United States, it remained legal elsewhere well into the nineteenth century. Slavery was legal in the Spanish Empire when Moses Austin received his initial contract to colonize in Texas, and it was legal in Mexico when his son Stephen had the contract reconfirmed in 1821.

During the 1820s, both national and Coahuila y Texas governments began a halting attack on the institution. Between 1823 and 1829, national authorities decreed a prohibition of the foreign slave trade (1823); the emancipation of slave children under fourteen (1823); an extra grant of land to settlers who brought in large numbers of enslaved laborers (1823); a reconfirmation of the proslavery article in the national colonization law (1824); the abolition of the internal slave trade (1824); the abolition of slavery in Mexico (1829); and the subsequent exemption of Texas from the abolition decree (1829). The state government at Saltillo was equally erratic. Among its decrees were a six-month period during which slaves could be brought into the state (1824); a six-month sunset period for slave importations (1827); a post-nati emancipation law (1827); a law providing for the emancipation of 10 percent of the slaves on any estate undergoing sale or transfer (1827); and a contract labor law that allowed Anglo slave owners to sign their bondsmen to ninety-nine year indentures, essentially undercutting all previous antislavery legislation (1827).[15]

Mexican antislavery affected the institution in Texas in two important ways. First, it made slave owners nervous, leading many to entertain ideas of derailing the movement either through skillful manipulation of the federal system or, if that failed, through other measures. By the 1830s, fears for the future of slavery were compounded with other political fears, prompting many Anglos to seek a solution through the drawing of a sovereign boundary between Mexico and Coahuila, and Texas—that is, through secession and political independence. The more they spoke of such things, the more they brought on the second development: the tendency of slaves to view any potential border as separating slavery from freedom. Before the 1830s, when the boundary between the United States and Mexico was at the Sabine River and it was far from clear that slavery was illegal in Mexico, slaves did not necessarily associate the border with freedom. After independence in 1836, with slavery firmly established in Texas and definitively abolished in Mexico, slaves built the border into an unequivocal boundary between slavery and freedom.[16]

The result was a steady flow of runaways over the 350 miles of dry, difficult terrain between the Brazos and the Rio Grande. The number of slaves who managed to survive the journey was well into the thousands.[17] On an 1854 visit to Piedras Negras, a small town on the Mexican side of the river, Frederick Law Olmsted reported seeing several former Texas slaves. One, a native of Virginia who had come to Texas with his master, spoke Spanish fluently, was a member of the Catholic Church, and had traveled throughout northern Mexico. The fugitive estimated that forty slaves had come through the town in the preceding three months. Some of the former slaves scattered and married into local families, but Olmsted also reported hearing of a community of runaways settling a few days outside Piedras Negras and comprising a virtual maroon colony. "The Mexican Government was very just to them," Olmsted quoted his informant, adding, "They could always have their rights as fully protected as if they were Mexicans born."[18]

In a simple economic sense, the loss of several thousand slaves was felt on the ledgers of the state's slave owners. Estimating the number of runaways is difficult. The U.S. census attempted an estimate, but the vast majority of census tracts reported no fugitives at all, a very unlikely circumstance. However, in what was clearly a departure from the practice virtually everywhere else in the South, the U.S. census-taker for Chappell Hill in 1860 reported the number of slaves "fugitive from the state" for each slaveholding. Out of a population of 1,984 slaves, 297, or 14 percent of Chappell Hill slaves were counted as fugitives. Why the Chappell Hill census was done differently is difficult to say. For reasons we can only guess at, the census-taker seems to have interpreted or asked the question differently. Most likely, he recorded the total number of slaves who ran away during the year, regardless of how long they remained away or where they went. If so, it is possible that some slaves were double-counted for multiple attempts.[19] Another contemporary estimated that four thousand slaves had successfully fled to Mexico by 1855, an aggregate loss of approximately 3.3 percent of the slave population.[20] If the true number of successful escapees was only a fraction of that, it would still represent a noticeable proportion of the labor force. But monetary costs do not tell the entire story. The threat of slave flight to Mexico and the costs associated with it worked their way into the masters' consciousness. When given the chance to swap a tract of land for a slave, Brazoria County planter Guy M. Bryan was forced to think carefully about the matter. "The negroe he has got Mexico in his head," he wrote to his brother-in-law, "on this account I may not buy." Other planters undoubtedly shared Bryan's apprehensions and factored the possibility of flight into their economic calculations.[21]

Flight was not the only form of border-oriented resistance. On at least one occasion, the slaves' vision of Mexico led them to abandon negotiation and attempt an overthrow of the slave system. In October 1835, a Mexican Army advanced toward the Austin Colony to suppress what had then become an open rebellion led by Anglo settlers. An unknown number of slaves near Brazoria, seeing an ally in the force that approached, rose up, only to be crushed. According to the one and only source on the rebellion, the rebels "had devided all the cotton farms and they intended to ship the cotton to New Orleans and make the white men serve them in turn." Almost a hundred slaves were hanged or "whipd nearly to death."[22]

Two decades later, authorities discovered what they said was another slave conspiracy. Whether it was real, a figment of Anglo imaginations, or some combination of actual plot and white paranoia is impossible to say, but the prominence of Mexico says something about its place in Brazos residents' consciousness. According to newspaper reports, two hundred slaves in Colorado County, just west of the Brazos counties of Austin and Fort Bend, planned to kill all the white men and "make wives" of the women. A search reportedly turned up a stockpile of pistols, long guns, ammunition, and bowie knives. A vigilance committee hanged the three men accused of being the ringleaders, while the lives of the rest were spared. As with so many other incidents, the border figured prominently. According to county officials, the slaves had resolved to "fight their way to a 'free state' (Mexico)." In addition, officials claimed that every Mexican in the county, "without exception," was involved in the plot. One man in particular, known only as Frank, was reputed to be one of the leaders, prompting officials to conclude "that the lower class of the Mexican population are incendiaries in any country where slaves are held, and should be dealt with accordingly." This sentiment translated into the expulsion of all Mexicans from Colorado and several nearby counties.[23] Although it is difficult to say for certain how much of the plot was real and how much was a figment of Anglo imaginations, it demonstrates the continuing linkage of Mexico and antislavery in master-slave discourse. Thus as Brazos masters and slaves spoke to one another, they did so in a paternalist idiom that was informed by a border: the awareness that the reach of the slaveocracy ended 350 miles westward. Mexico was perpetually in the heads of both masters and slaves.

Given the specific cultural inputs of paternalism—most notably evangelical Protestantism—it seems unlikely that African-born slaves would negotiate as creoles did. However, Africans appear to have learned the language of paternalism quite well, at least as quickly as they learned to communicate in English. In 1844 a Brazoria County slave named Garbo apparently persuaded his owner to let him find a new master. Garbo apparently had an abroad wife who was owned by an S. M. Swenson, and he seems to have been sent out by himself to Swenson's neighborhood to see if he could find a planter willing to buy him. In the end he did not succeed, but the amount of freedom he was given is testimony to considerable persuasive ability.[24]

The Africans' rapid adaptation to conditions in Texas does not imply in any way an acceptance of slavery, any more than such adaptation did for creoles. To the contrary, the woods and canebrakes of Brazoria County seem to have abounded with African fugitives, some of whom succeeded in hiding out for long periods of time. The newspapers of the 1830s and 1840s contain several advertisements for African-born runaways, including a few who were headed "westward," presumably toward Mexico.[25] But most Africans quickly assessed their situation and responded accordingly. Virtually all of the Africans on the Brazos would have been familiar with African versions of slavery, whether as a slave, as a member of a slave-owning household, or simply having lived in a slave-owning society. Although the specific cultural associations of American slavery would have taken time to translate, the essential negotiation between master and slave would have been familiar. Undoubtedly, Africans drew on a familiarity with master-slave negotiation, which helps account for their astuteness.

"Axe, Plough, and Hoe"

"Happiness might be promoted," Stephen F. Austin once wrote to his cousin Mary Austin Holley, "by conquering a wilderness by the *axe*, the plough, and the hoe." His reference to a wilderness speaks for itself, but at the same time his phrase speaks to another, equally important aspect of the master-slave relationship—work. The Brazos, after all, was settled when it was and the way it was by people who wanted to profit from the market for staples like cotton and sugar, and from the start, most understood that the bulk of the work would be done by slaves. Work, as historians Ira Berlin and Philip Morgan point out, "necessarily engaged most slaves, most of the time" and constituted "both the source of the slaves' oppression and a seed of liberation."[26] Between these two poles lay a vast, negotiated middle ground. For the most part, the transcripts of these negotiations do not survive. We never actually hear slaves threaten to run to Mexico if their desires are not met, but then again, such threats were as a rule implicit rather than explicit. We can, however, see traces of these negotiations through a close examination of the region's labor routines, the object of so much of the daily "war for position" waged perpetually by masters and slaves. Control over the conditions of labor, material goods, and community time were the currency of these negotiations, all carried out against the backdrop of wilderness conquest and border flight.

More than most other plantation regions, work along the Brazos was tremendously varied. The triplex of "axe," "plough," and "hoe" actually speaks to one aspect of this variability—change over time. As suggested by the axe, the initial labors on almost all Brazos farms and plantations revolved around clearing land, although, as we shall see, the axe was not always the principal tool. Beginning in the river bottoms of the Lower Country in the 1820s and 1830s, spreading gradually into the bottoms of the Upper Country in the 1840s, and eventually into areas outside the bottoms in the 1850s, land-clearing put its stamp on the master-slave relationship in the earliest phase of local settlement. Austin's reference to "plough and hoe" spoke to agriculture, which gradually supplanted land-clearing as the dominant task. Yet agricultural work routines varied greatly. Early on, farms and plantations were careful to balance their staple crop production with food production and stock raising, and cotton eventually edged out the former two activities to dominate Brazos work routines. Beginning in the 1840s, however, sugarcane began to catch on in the Lower Country, especially in Brazoria, introducing yet another variation in the region's labor regime. Each crop routine presented slaves with different demands, as well as with different openings and opportunities for autonomy. Work, then, contributed substantially to the mosaic that was Brazos slave society.

For virtually all early newcomers, one of the most powerful symbols of the Brazos's enduring wilderness status was its vast expanse of uncleared land. Most migrants hailed from places where the land had been ruled off, with greater or less thoroughness, into cultivatable tracts. Even those who came from more recently settled states like Alabama and Mississippi, where travelers often noted the slovenly appearance of new and backwoods farms, viewed the Brazos as virgin land in need of some degree of preparation before coming under the plow. Slave laborers cleared most of the land along the Brazos, and land-clearing, as an important element of the region's early labor regime, exerted a considerable shaping influence on slave life. In cases where the master and his family shared the task of clearing land with a small number of slaves, a process that might continue for several seasons, a close "sawbuck equality" may have resulted. However, land-clearing did not always take place under such classic "pioneer" conditions. Variations in topography meant that the actual work differed from locality to locality, easier in some places than others. Large gangs of slaves working under the direction of the master or an overseer were probably responsible for clearing most of the Brazos bottoms, a situation less conducive to tight, mutually dependent relations between master and slave than to community bonds among slaves. Finally, even those plantations with large amounts of cleared land often required more, whether to expand production or to replace exhausted soils. Clearing land was thus not limited to the earliest phase of settlement; it was a continual, if gradually diminishing, facet of slave life, superseded inevitably by a focus on agriculture.

In the Lower Country, especially in the Brazos and Caney deltas, the effort required to prepare land for cultivation varied considerably, depending on the topography. Early settlers recognized three basic soils: cane land, prairie land, and bottomland. The most prized of the three was cane land, also known as "peach and cane land" for the wild peach trees that sometimes dotted the expanses of canebrake.[27] Such lands were quickly and easily cleared. Simply setting fire to the cane, as a number of travelers did for sport, could clear several acres in just a few days.[28] Surviving trees could either be allowed to stand or could be removed by girdling (chopping a ring in the bark, killing the tree slowly by choking off its vascular system), burning the stump, and pulling it out with horses or mules. Fallen logs would be gathered together (the communal labor ritual known as "logrolling") and burned, providing the soil with a natural fertilizer. Sometimes the process was followed to completion, but other times girdled trees and burned stumps might be allowed to stand and rot for many seasons.[29] In any event, land-clearing in the canebrakes was not always an arduous task. It was often quick and easy, a factor that contributed to the Lower Country's faster development and also meant that land-clearing did not necessarily entail extended periods of shared labor or result in tighter relations between master and slave.

The prairie lands outside the canebrakes and bottoms of both the Upper Country and the Lower Country were also relatively easily cleared. Brazos planters, like most nineteenth-century Americans, generally viewed prairie land as less desirable than the other varieties. Timber was a sign of fertility; lack of timber suggested poor soils. The prairies' reputation, along with the relative distance from river transportation, meant that land prices were lower, attracting greater numbers of nonslaveholders, who used unbroken prairie land as pasture. White family labor was responsible for preparing much of the region's prairie lands for cultivation, a moderately easy task, at least when compared with the wooded bottomlands. Sometimes landowners prepared the prairies by leasing land to newcomers and requiring a certain amount of "improvement" as part of the rent. As plantations expanded onto the prairies in the 1850s, slaves quickly cleared remaining lands. As with cane land, the clearing of prairie land could be done without slave labor, and thus did not necessarily foster close mutual dependency between master and bondsmen.

However, clearing the bottomlands outside the Brazos delta and in the Upper Country was arduous work indeed. It took several decades to transform low-lying land where, as one traveler put it, "tree trunks and underbrush had grown into a massively thick, matted, and at times thorny carpet covered by every type of tendril and vine" into cotton and canefields.[30] Slaves were still clearing Upper Country bottomlands right up to the Civil War. If any environment required sustained, cooperative labor to prepare for the plow, it was the bottomlands of the Brazos, San Bernard, and Navasota rivers, along with Oyster and New Year's Creeks. But the clearing of this land was not accomplished by masters working alongside small groups of slaves; it was done by large gangs of slaves, a process that encouraged community bonds among slaves themselves more than sawbuck equality. Elsewhere in the Old Southwest, where communications with the seaboard states were better, the initial clearing of land might be done by an advance party, but in Texas during the 1820s and 1830s, poorer communications with the United States meant that early planters often arrived with entire slave labor forces. Or, if an advance party was sent, the bulk of the labor force usually arrived shortly afterward. Jared Groce, the region's first planter, arrived with all ninety or so of his slaves in 1822. According to a census conducted in 1825, of the eleven masters with more than ten slaves living in the Austin Colony, nine had all of their slaves with them. The remaining two had each brought four slaves, presumably to begin clearing and cultivation, and were noted in the census to be sending for their remaining laborers.[31]

With slaveholders preferring to bring entire contingents of laborers with them to clear large amounts of land, slaves often outnumbered whites. Of necessity, some masters conceded a fair measure of autonomy to their slaves, recognizing that close supervision of a large number of laborers might cut into the amount of land cleared. On a visit to Texas in 1832, Quaker abolitionist Benjamin Lundy found the laborers at William Stafford's Lower Country plantation to be working with no white supervision at all. Lundy attributed this indulgence to Stafford's Quaker background, but it was likely that Stafford's slaves played an active role in negotiating the situation. In another case dating to 1838, a slave named Bill was left in charge of a farm along the San Bernard during what was likely the land-clearing phase while his master traveled on an errand. The owner, Arthur Hennis, in a near-perfect demonstration of borderland paternalism, instructed Bill to call on a neighbor (who actually lived several miles away) "for protection in case of any difficulty."[32]

The frequent unsupervised activity of slaves during the land-clearing phase was also evident in the built environment. To be certain, owners of very few slaves might share a cabin with them, but those who owned more than three or four slaves generally housed them in outlying buildings. Brazoria's Edwin Waller lived in what one visitor called an "an exceedingly rude, open, double log cabin, with two shed rooms" and housed his seventeen slaves elsewhere on the grounds. Others took advantage of the same ubiquitous two-room dogtrot-style cabin, in which two small log pens spaced a few feet apart were joined by a common roof to create a breezeway between them, to separate whites from blacks. Ann Coleman, who traveled through Brazoria in 1832, stayed with a planter who used one room for his own family and the other as a kitchen and "house" servants' quarters. The remainder of his thirty slaves crowded into just a few outbuildings. "I had to sleep with the [master] and his wife," she recalled, "I slept at the back of the bed, the wife in the middle, and the man in the front."[33]

As the plantations boomed in 1850s, the landscape was transformed. In 1860, the "improved" acreage of the five Brazos counties was nearly ten times what it had been in 1850, with the most spectacular gains coming in the Upper Country counties of Austin, Washington, and Grimes. Because slaves did most of the work, and often without sustained supervision from whites, it is not surprising that they saw themselves and not their masters as wilderness conquerors and community builders. Land-clearing, however, persisted as only a secondary component of the labor regime once enough was done to allow agriculture to take hold.[34]

While laborers gradually tamed the Brazos wilderness, they simultaneously produced crops for both markets and sustenance. Planters (and very likely slaves) were mindful of the need to coordinate the timing of land-clearing with crop production. Newcomers often aimed to arrive during the early spring. After clearing a small field, they planted the requisite crops. Once a place was up and running, planters took advantage of the gap between fall harvest and spring planting to clear additional acres. Land-clearing and agriculture, then, were continual and complementary activities, the labor requirements of each influencing slave life and culture.

For nearly all slaves in the 1820s and 1830s, agricultural labor involved cultivating cotton and corn, along with lesser crops like peas and sweet potatoes in between times. New arrivals generally concentrated on growing corn and raising stock, adding cotton only when the farm was able to produce a basic level of sustenance. In most cases, this exclusive focus on food crops was short. As early as 1824, Stephen F. Austin accepted—and clearly expected to receive—payment of land fees in the form of cotton notes, suggesting that cotton was widely cultivated by then.[35] Total production in these early years, however, was low, reflecting settlers' "safety-first" approach. One contemporary estimated the 1828 Brazos crop at 500 bales, a figure that, according to one Mexican official, jumped to 5,000 bales four years later.[36]

For most early Texas slaves, who migrated from cotton-producing areas, the work routine was nothing new. For a significant minority, however, who came from noncotton states, the monotony of cotton work likely required an adjustment. Slaves from Upper South states like Kentucky and Missouri, who happened to make up a significant proportion of the early population of Gulf Prairie, were probably accustomed to tobacco work, which was usually organized into relatively small squads. Cotton planters, in contrast, relied on much larger gangs and were more likely to work them under an overseer, leaving fewer opportunities for skilled or leadership positions. The experience of slaves from noncotton states reflected the more diverse Upper South economy. For example, before coming to the Brazos, James F. Perry's slaves had spent time working the family's lead mines. The switch to cotton for these slaves entailed a leveling process as well as the imposition of a tedious, sunup-to-sundown routine.[37]

Although cotton routine had changed somewhat during the antebellum period, growing more efficient through the use of more sophisticated tools and more cost-effective through greater attention to food crops, the basic work calendar along the Brazos had changed little when Fort Bend County planter Jonathan Waters described it in a letter to *De Bow's Review* in 1859. Waters was an exceptionally wealthy planter, the owner of more than two hundred slaves in 1860, so his routine would inevitably have been simplified by the less well-to-do, but in its outline it probably represents the norm for most planters. Shortly after the new year began, slaves prepared for planting by using a two-horse plow to throw up beds spaced six to eight feet apart. Because of Texas's low latitude and hot climate, planting began around March 10, about two weeks earlier than in Natchez, Mississippi. Using a special plow, gangs opened up the beds to a depth of four to six inches and threw handfuls of seed into the groove. Another pass with a harrow, an implement resembling a horse-drawn rake, buried the seed and broke up dirt clods. Waters then had his slaves pass a roller over the beds to pack down the earth. Less elaborate operations used simpler tools, like wooden blocks or logs. From the time the sprouts appeared to midsummer, workers "scraped" the cotton, or weeded it using hoes, and thinned out the row to leave two feet between plants. Meanwhile, plow gangs would pass down the rows, loosening the soil and killing grass and weeds.[38]

This phase of the cotton routine changed slightly from the earliest days of settlement to the eve of the Civil War. On the earliest cotton farms and plantations, planters preferred to use male slaves to operate the plows and sweeps that prepared the soil. The relatively small size of early slaveholdings and the strategy pursued by some planters that involved buying extra female slaves to produce future laborers meant that many masters depended on women to do plow work. Although the use of women on plow gangs presumably diminished as Brazos slave owners increased their holdings, especially in the bottoms and in the Lower Country, women plowed on some of the smaller Upper Country farms through the antebellum period. In addition, many Brazos plow gangs worked under the direction of a master, rather than under a black driver. Work under a master or an overseer tended to be the rule in cotton, whereas in tobacco foremen led smaller gangs and in rice a task system prevailed. Even on the largest plantations, where separate gangs of enslaved workers might toil on far-flung quarters, masters preferred not to rely on slaves to supervise other slaves. Thomas Blackshear used his sons to work the noncontiguous tracts that made up his Grimes County plantation. The only black driver on his plantation supervised the "trash gang," composed of children and less-productive adults. The work of the planting season, then, tended to level gender hierarchies on smaller holdings even as it reinforced them on larger ones, all the while subjecting all slaves to increased scrutiny from the masters.[39]

By late June the cotton and corn stalks were tall enough to shade the ground and prevent weeds from taking over. All that would be required now was a small amount of hoeing. This was the time of year known as "lay by," a time devoted to chores other than cotton and corn work. Some farmers and planters used the time to work peas, sweet potatoes, and other food crops. July might also be given to pulling fodder, digging wells, or splitting fence rails. The first cotton blossoms, or "squares," appeared along the Brazos in mid-May. Bolls actually began to open in late July, but early pickings were slim. Owners of many slaves might not bother to send their best workers into the fields, instead using a trash gang to do light picking. As more bolls opened, planters quickly sent all capable hands into the fields.[40]

Picking cotton was skilled and painful work. Bolls had to be plucked quickly and without damage to the staple or the plant itself—broken branches produced no more cotton. Workers unaccustomed to cotton-picking, as many newcomers were, soon developed raw, bleeding fingers, as they raced to complete their tasks and avoid a date with the lash. Solomon Northup, who picked cotton in Louisiana during the 1840s, recalled his first day in the fields. "It was an awkward business indeed," he wrote. "While others used both hands, snatching the cotton and depositing it in the mouth of the sack, with a precisions and dexterity that was incomprehensible to me, I had to seize the boll with one hand, and deliberately draw out the white, gushing blossom with the other." At the end of the day he had picked only ninety-five of the expected two hundred pounds, a transgression for which he received a one-time pardon. Even veteran cotton hands suffered from back and joint pain as they strained to reach the bolls, a motion repeated from sunup to sundown, and all left the fields consumed with fatigue. To make matters worse, the late-summer Texas sun could be unbearable, especially when compounded by the humidity of the river bottoms, to the point where workers frequently fainted in the fields. Shade became a rare luxury as slaves cleared ever more acreage, transforming one-time timberlands into flat, treeless expanses of cotton. In addition to coping with these hardships, slaves needed to be careful not to pick "trash" along with their cotton. Leaves, stalks, and dirt mixed in with the fleece lowered the price, and slave owners exacted retribution for trashy cotton with the whip.[41]

For slaves, the same daily routine dominated the late-summer and fall months, Sundays and the rare rainy day excepted. Following a predawn meal, slaves made their way out to the fields, each carrying a large sack and a much larger basket. They dragged the sack as they picked, taking care not to allow it to damage cotton plants. Laborers generally moved between rows, picking from both left and right sides as they went. When the sack was full, they emptied it into their basket, which was usually placed at the end of a row, then packed the cotton down by stomping on it. At the end of the day, laborers carried the baskets to a storage or gin house. There began the nightly ritual of weighing and recording each person's pickings.[42]

Cotton-picking was "tasked," though not in the same way that rice was in the Carolina Lowcountry, where slaves who finished their work had the rest of the day to themselves. Rather, a cotton task was more akin to a quota, an amount that an individual hand was expected to pick. While tasking on Carolina rice plantations allowed slaves some control of their time, tasking in the cotton belt left control of time firmly in the hands of slave owners. Masters and overseers calibrated tasks individually according to their assessment of a hand's ability, and variation from worker to worker and season to season was great. Proper picking demanded dexterity more than brute strength, which meant that women could often gather as much as or more than men. Planters also had to temper their expectations according to the condition of the crop. A field that had just begun to open or one that had been picked over yielded less than one at the height of the season. For this reason, growers often attempted to stagger planting to allow different fields to open at different times. Poor weather might also reduce what a planter could reasonably expect from a slave (although reasonableness was never guaranteed), and the "cotton worm," or caterpillars, which began appearing on the Brazos annually starting in the early 1830s, also cut into yields. The standard punishment for failing to fulfill a task was whipping, and not even young children were exempt. Ann Raney Colman, who married a Brazoria slave owner in 1832, sent her six-year-old slave into the fields expecting her to pick fifty pounds of cotton per day. On several occasions the child picked for a short while, then ran away. When she returned, she faced a beating, not only for running away but for failing to gather her fifty pounds.[43]

The work of cotton-picking influenced the master-slave relationship in complex and contradictory ways. Cotton tasking had the potential to corrode community ties among slaves, the desire to demonstrate one's ability nurturing a competitive ethos. Contests to see who could pick the most cotton in a day were common throughout the South, with some of the more extraordinary totals receiving mention in the regional press. But this embryonic competitive tendency was more than offset by slaves' efforts to harmonize work with community goals. The task system did not lead inexorably to full-fledged individualistic competition. Slaves were well aware that competition ultimately worked against them, both at the individual and the community levels. Solomon Northup explained the calculus that was no doubt familiar to Brazos slaves as well: "A slave never approaches the gin-house with his basket of cotton but with fear. If it falls short in weight—if he has not performed the full task appointed him, he knows that he must suffer. And if he has exceeded it by ten or twenty pounds, in all probability his master will measure the next day's task accordingly."[44]

Slaves combated their masters' demands in a variety of ways. Some forms of resistance, like attempted flight, whether toward Mexico or simply to a local haven, were likely to result in only temporary relief followed by "correction." Some slaves, through the calculated feigning of illness or weakness, might succeed in persuading their masters to reduce their tasks. When such maneuvers worked, it was likely only after a long campaign, punctuated by repeated punishments. The net effect of these actions, though not necessarily conscious, appears to have been an array of concessions granted by masters to their slaves. Although the source base is narrow, it appears that Brazos masters gave ground on fronts ranging from control over time and community life to material conditions. Little documentation exists for the actual process of negotiation. More often we hear the echoes of earlier master-slave conflicts, manifested in what proslavery ideologues liked to portray as paternalistic indulgence. Take for example the slaves at Peach Point Plantation, who were allowed to pick cotton on their own time and accumulate credits on the plantation ledger. During the late 1830s and early 1840s, at a time when cotton prices hovered at around six cents per pound, Peach Point slaves earned ten cents a pound, suggesting the operation of a negotiated moral economy. A few slaves, such as Peter, Simon, and Sam, accumulated $30, $40, or even $50 in credits, which they spent on such goods as tobacco, fabric, shoes, and even a dress pattern from New Orleans. And although men probably had more access to money than women, some women also accumulated respectable sums. A Brazoria woman named Elizabeth managed to purchase (through her master) two cows "with her own money, in slavery." Over time the cattle increased "considerably," but upon the death of her master, her lack of legal title allowed the estate executor to sell the animals to a neighboring planter. After the war, Elizabeth enlisted the aid of the Freedmen's Bureau in recovering the cattle, but whether she succeeded or not is unknown. Her experience serves to highlight the precarious nature of all master-slave negotiation, which could be undone at a moment's notice by the caprices of a master or an executor.[45]

James F. Perry also allowed slaves to complete each other's picking tasks, essentially converting individual task expectations to a single communal one. Perry kept close track of how much cotton each slave picked for the other. "George pick for Simon 191 [pounds]," ran a typical entry in the plantation ledger for the 1845 season. All of these exchanges appear to have involved people living in separate households and not nuclear family members. As we saw in chapter 2, both known Africans at Peach Point, John and Bill, took part in these exchanges, which raises the possibility that some understanding of fictive kinship, rooted in West African experience as well as New World pragmatism, played a role. To a considerable extent, then, slaves on the larger plantations were able to break down whatever individualizing tendencies existed in task work and transform it into an agent of community.[46]

Community time, including the chance to travel to town to attend dances and other events, was one of the most coveted prizes for slaves and the object of much negotiation. Rebecca Pilsbury, a slaveholder's wife and a northerner who was unaccustomed to supervising unfree laborers and had to turn to a neighbor for help, gave in on a number fronts when her husband, a U.S. representative, attended sessions of Congress. She allowed their slaves to work independently and attend Christmas events at Brazoria on their own. In 1852 slaves in the neighborhood of Washington-on-the-Brazos held a large dance on the edge of town "by permission." When the dance was violently broken up by a white man, the slaves "came into the town and made known what had transpired, asking at the same time, the protection of their masters and others in their enjoyment of the dance. . . . On the receipt of this intelligence, quite a number of respectable citizens repaired to the spot, for the purpose of protecting the negroes in the enjoyment of their authorized amusement." Such privileges were granted not out of masters' benevolence; they were hard won through slaves' willingness to malinger, flee to the woods, or make a try for the border.[47]

Community time was one of the most valued of all negotiated commodities, but the desire of slave owners to turn a profit imposed very real constraints. The amount and frequency of free time—and hence the texture of community life—was governed to a large extent by the needs of the cotton and cane culture. Not only did each crop have a different calendar; the labor demands themselves influenced community life in complex ways. Some tasks were done under the direction of the master or overseer, which placed an emphasis on the bilateral relationship between the master or overseer and individual slaves. Other routines were more cooperative in nature, fostering lateral ties among members of the plantation workforce.

Whether cotton routines fostered a close master-slave relationship or tight-knit community bonds among slaves depended largely on the size of the operation. The majority of Brazos slave owners in the 1830s owned from one to five slaves and worked fairly closely with them. But we should be careful not to overdraw the portrait of shared labor and tight bonds on small farms. It did not take very many workers to create a situation in which the head of the household became more of a foreman than a worker himself. While we lack good records for small farms on the Brazos in the 1820s and 1830s, the one operated by William Bradbury in 1859 and 1860 in Austin County was probably similar to earlier ones. Bradbury relied primarily on his two sons' work year-round and supplemented their labor during planting and harvest seasons with a free white laborer, and sometimes up to three. In addition, he hired a single slave woman each January 1. Diary entries reveal Bradbury's involvement with the work performed on his farm but not sustained participation in it. Whatever actual labor he did seems to have been interspersed with monitoring work on other parts of his farm. "Alfred and Andrew hoed corn to day until after dinner," read a typical diary entry, suggesting a certain managerial detachment. Patsy and Nance, the slave women he hired in 1859 and 1860, respectively, seem not to have worked much in the fields and largely escaped his supervision, though likely not that of Mrs. Bradbury.[48]

While some slaves on small farms in the 1820s and 1830s worked closely with their masters, it was not uncommon for others to work alongside hired white laborers. According to the Austin Colony census of 1826, thirty-four of ninety slave owners also employed free white laborers. Listed as "laborers or renters," they almost certainly included a few overseers, as well as artisans like Noah Smithwick, who worked as a blacksmith on the McNeel plantation in the mid-1820s.[49] Others almost certainly worked more closely with slaves. John P. Coles, a close ally of Austin's in the Upper Country, housed five white laborers in addition to his four slaves and five family members. But such situations seem more often to have produced animosity than cooperation between slaves and free laborers, as whites sought to assert themselves as de facto foremen. In 1838, a slave named Charles stood trial for murdering Alexander Carmack, a hired man on Isaac Donohoe's Austin County farm. The two had been feuding for two weeks, ever since Carmack had threatened to whip Charles for "hollowing and singing" as they pulled fodder with three other hired men and three slaves. Charles's smoldering animosity finally erupted when Carmack ordered him to fetch some watermelon and then forced him to carry his (Carmack's) rifle as they walked along a road, alone. Furious over Carmack's demands, Charles shot him in the back of the head and ran away. He returned fifteen minutes later and dragged the body away from the road, where it was picked clean by buzzards within the week. Charles was forced to confess weeks later when other slaves told Donohoe that Charles had offered

Carmack's shoes for sale and some of his shirts as a gift. Not all interactions between slaves and free laborers ended in violence, but they were undoubtedly fraught with similar tension.[50]

If the small farm typified most masters' experience with slavery, larger farms and plantations typified most Brazos slaves' experience with their masters, even in the early years. As early as 1826, 61 percent of the slaves in the Austin Colony (the vast majority of whom lived along the Brazos) lived on holdings having more than ten slaves. They were also concentrated in an area that eventually became Brazoria and Fort Bend counties. A little more than ten years later, over three-quarters of Brazoria County slaves worked for masters who owned more than ten slaves, and 57 percent lived on holdings with more than twenty.[51] Paradoxically, then, while most masters experienced shared labor with their slaves on small, isolated farms, most slaves, even during the Mexican period, experienced communal labor alongside other slaves and under the master's or an overseer's supervision.

Beginning in the 1840s with the spread of sugarcane cultivation in the Lower Country and accelerating rapidly in the 1850s, the Brazos labor regime grew more variegated. The already-sizable cotton plantations of Brazoria County and a part of Fort Bend County became even larger, reflecting the economies of scale associated with sugar. For several thousand enslaved Brazos residents whose lives had been structured by the demands of cotton, cane cultivation brought significant, mostly unwelcome changes. The demands of the crop were notorious; slaves throughout the South dreaded being sold to a Louisiana sugar planter. Those slaves who lived along the Brazos had been lucky enough to escape being sent to Louisiana, but, unfortunately, sugar came to them.

Most Brazos sugar plantations were converted cotton plantations. Cotton cultivation had entailed a deskilling of work and a leveling of the occupational hierarchy, which, for people accustomed to crops like tobacco and rice or those accustomed to life in urban areas, required adjustment. Very few Brazos slaves came from Louisiana, so almost none were veterans of sugar work. The group with the most sugar experience was probably the African-born population of Brazoria County, an undetermined fraction of whom had spent time on Cuban ingenios. Even there almost all evidence points to a very short stay in Cuba and correspondingly limited sugar experience.[52] To a certain extent, cane cultivation reversed the leveling trend associated with cotton by introducing more skilled work, some of which carried extra payments as an incentive.

These privileges were a welcome perquisite for those (usually male) slaves who were in a position to benefit. However, these minor concessions did little to mitigate the overall harshness of the work. To make matters worse as well as more complicated, several Lower Country planters grew both cotton and cane, resulting in a grueling, hybrid work regime with precious few slow periods.[53]

The cane calendar began at roughly the same time and in the same manner as cotton's, with plow gangs turning over the soil and spacing furrows two and a half to four feet apart. When it came time to plant a new field, hands, mostly women and children, took stalks of cane from a cart and placed them length-wise in the furrows. Trailing the cane-droppers were prime male and female hands with cane knives, which they used to cut each stalk into three pieces. Planters' publications suggested they do so because the strongest shoots grew out of either end of a piece of cane. Finally, trailing the cutters came a turning plow, which threw dirt back on top of the cane. These plow gangs passed through multiple times to be certain that each stalk lay under at least six inches of soil. For those slaves who also worked cotton, this phase of cultivation was not a serious departure from familiar routines. Hands simply moved from cotton field to cane field and back and worked in gangs under the direction of the master or overseer.[54]

The work routine remained essentially the same as cotton when the first shoots began to appear, although, as Fort Bend County planter Jonathan Waters counseled, "Cane requires a higher degree of cultivation than corn or cotton." What he meant was that it needed to be worked more frequently. Once a field was full of cane sprouts, hoe gangs came through and scraped, or loosened the soil until the original seed cane lay about three inches below ground. The process had to be repeated after every good rain to prevent the soil from hardening and making it difficult for young shoots to emerge. By April, the young cane required plowing, first a pass to throw dirt away from the row, and ten days later another pass to throw the dirt back toward it. In between plowings, hoe gangs weeded, kept the soil loose, and destroyed any new shoots that popped up, lest they sap the growth of the others. Brazos cane could be laid by around June 20, although slaves occasionally had to kill off troublesome tie vines, which plagued the bottomland plantations. By this time of year the cane was quite tall, inspiring at least one Brazos resident to wax poetic on her grandfather's "cane field with its lighter green blades, swaying and rustling with every passing breeze."[55]

It was during the laid-by period that the cane and cotton routines diverged. On all plantations, this relatively slow time was used for a host of tasks, including work in food crops such as corn and potatoes and sundry chores such as fence-building and basket-making. Enslaved sugar workers performed the same tasks, but they also needed to chop wood to fuel the steam-powered mills that pressed the cane on most Brazos plantations. Wood-chopping was tasked in a manner similar to cotton-picking: each hand was expected to produce a certain daily amount. As with cotton tasks, failure to meet a fuel quota was punishable by whipping. During the summer of 1859, Stephen S. Perry of Gulf Prairie found himself frustrated at the recalcitrance of William, a hired slave. According to Perry, William had "failed for several days to cut his task of wood one and a half[,] The same I give my other men." Because William belonged to another, Perry had shown restraint and not whipped, but he wanted William's owner to know that "he deserved to be flogged and if he had belonged to me I would have whipped him the first day that he failed to get his tasks not having cut wood."[56]

Rolling season, or grinding season, began on the Brazos in mid-October, and it was at this point where cotton and sugar routines diverged most sharply. From now through December or even January, Lower Country sugar plantations ran at a brutal pace, one set by the mill. Sugarcane's vulnerability to cold weather added a sense of urgency to the proceedings. One good, sustained blast of Arctic air from Canada could jeopardize an entire crop, especially if it arrived before seed cane had been cut and stored, which was always the first priority for the season. At the first sign of cold weather, planters halted production and sent slaves into the fields to "windrow" the cane, or cut it and stack it so that the outer layers protected the inner layers from frost. In 1856 a series of "Northers" killed the entire Brazos crop before the seed cane could be cut. The loss of the seed and a drought the following year meant that it took planters until 1858 to revive production. Frost threatened cotton as well, but because of the plant's hardiness, combined with the seeds' resistance to cold weather, the risk of losing an entire crop was much lower than with cane. The fear of damage prompted Brazos sugar planters to drive their hands especially hard during the harvest.

Records for rolling season along the Brazos are sparse, and those that survive reveal little about the organization of work. Estate inventories afford us a glimpse of a revived occupational hierarchy that cotton had wiped out. One inventory from the Abner Jackson plantation, across the Brazos from Gulf

Prairie, listed a carpenter, a blacksmith, and a brick mason. Like many sugar planters, Jackson also used slave coopers, in this case two men, to supply the plantation with hogsheads for sugar and smaller barrels for molasses. Many of these artisans likely received payments of some type. During the year 1860, Simon Neece, the slave cooper at Peach Point Plantation, received 85¢ for each hogshead he built, 50¢ for each he repaired, 6¢ for each stave he jointed, and 75¢ for each barrelhead he made. Neece received payments of $13 in cash, $15 in molasses (approximately fifty gallons, suggesting it may have been intended for resale), and the payment of notes worth $4.25 and $39.50. A few (though apparently not most) sugar planters entrusted their most expensive piece of equipment—the steam-powered mill—to slaves: Abner Jackson's inventory lists a man named Big William as a slave engineer. Assistants to the engineer were also common, although their skills were not always noted in inventories.[57]

The sugar-making process in Texas is very poorly documented, but we can gain an idea of its workings by examining it for southern Louisiana, where the climate and the soils were similar, allowing for the employment of similar techniques.[58] The task sheet for a plantation owned by John Randolph of Iberville Parish, Louisiana (who later moved to the Brazos), reveals that sugar required a higher degree of coordination and specialization among workers than cotton. The largest single contingent of laborers was the cane-cutters, consisting of thirteen men and seventeen women. At the start of the harvest, gangs of cutters moved through the fields wielding machete-sized cane knives. Cutters aimed the blade at the bottom of the cane and severed the stalk in two or more withering blows. The lead cutters then moved on to cut more cane while trailing cutters hacked the leaves off the stalks and stacked them high on a cart. The Randolph plantation employed eight carters to carry the cane to the mill. All of these were men, indicative of the privilege that likely attached to the job; they were not cutting cane. At least one Brazos plantation, again at Abner Jackson's Lake Jackson, actually had a small railroad to convey cane to the mill. Cutters worked in gangs from "sunup to sundown," and beyond if the crop was threatened. Night work in the fields, however, was not common. A reasonably sized group of cane-cutters could chop enough cane in a day to keep the mill working all night. In sugar, the production bottleneck was in the rolling and processing, not the actual harvest.[59]

Waiting at the mill were between twelve and twenty hands whose job was to unload the cane and place it on the cane carrier, a conveyor belt that brought the cane from the mill yard up to the second or even third story of the mill, where the rollers were. This work continued around the clock in three different watches. Many of the cane-cutters were required to take an extra watch, which meant that they worked roughly eighteen hours in a twenty-four hour period, not including work done for themselves. When the cane reached the mill room, a hand designated as the "receiver" fed the cane into the mill by hand. At Randolph's Louisiana plantation, one man worked two watches and was spelled for the third by a cane-cutter working his second shift. The long hours led to dangerous bouts of fatigue. Occasionally, the receiver might catch an arm in one of the rollers, becoming maimed in a manner reminiscent of an industrial accident. Receivers fed the cane through at least twice, squeezing out as much juice as possible. At a few mills, the carrier fed the cane directly into a hopper, which in turn fed it into the mill, obviating the need for a receiver.[60]

The juice ran down a spout, through a sieve or piece of gauze, and into a large vat. Once enough juice collected and the solid matter settled to the bottom, the juice was brought into the boiling room for further processing in open kettles. Each kettle was progressively smaller than the first, and their French names—*grande, flambeau, sirop,* and *batterie*—survived in Texas as reminders of the process's Louisiana and French Caribbean roots. With a controlled fire roaring beneath, a sugar-maker, often a free wage laborer from Louisiana, supervised the boiling of the syrup. Each step from *grande* to *batterie* removed more impurities. At the final stage, the sugar-maker added lime to the syrup to induce a "strike," or crystallization, the moment of truth for a sugar-maker. After a successful strike, the sugar was placed into hogsheads that held one thousand pounds or more with drainage holes in the bottom, and the vessels were stored in a section of the sugarhouse known as the *purgerie.* Smaller barrels were then placed beneath the hogsheads to catch the molasses.[61]

Work in the sugar house, while grueling, demanded a respectable amount of skill. In addition to the engineer and his assistant who tended the machinery itself, a full-time fireman was required to maintain the correct steam pressure, as on a locomotive. At Randolph's operation in Iberville Parish, this job belonged to three men, each of whom worked a shift as a fireman in addition to shifts as carters and kettle hands. One slave per watch was responsible for maintaining the fire in the flue underneath the kettles. The kettle firemen worked their second watches as cane carrier hands. A group of four to five slaves, all men, worked around the clock in three separate watches as kettle hands, skimming impurities and ladling juice from *le grande* down to *le batterie*. Although Randolph entrusted this task to male slaves only, he did not apparently regard it as a skill on a par with some others. Several who worked the kettles at his place worked second watches as cane-cutters and carrier hands, although two of them worked consecutive watches at the kettles. To keep the steam mill and the kettle fires roaring, Randolph employed a full-time wood-hauler. A cane-knife-grinder, a cart-greaser, a cook, a cook's assistant, and a lamp-cleaner (indicative of the twenty-four-hour work routine) rounded out the rolling-season labor force.[62]

Rolling season, with its need for skilled and semiskilled work, reintroduced occupational hierarchy to the slave labor forces of the Brazos. Because they needed coopers, blacksmiths, engineers, firemen, and kettle hands, planters found it expedient to offer new privileges, often in the form of payments, so that operations would run smoothly. Thus it appears that sugar work put more cash into the hands of slaves than cotton did, and the fact that most of those in privileged positions were men meant that the benefits tended to reinforce a gendered hierarchy in the quarters. However, the hierarchy and privilege created by sugar work was unstable and impermanent. All but the infirm worked two watches, so that even in a given twenty-four-hour period, few slaves worked the same skill position the entire day. Those who worked as carters, firemen, and kettle hands by day toiled as carrier hands by night, alongside the cane-cutters. Privileged slaves would certainly have been aware of the possibility of reassignment. In addition, rolling season spanned at most one-fourth of the calendar. For the rest of the year, many positions of privilege ceased to exist, although some, such as the coopers, worked primarily outside of harvest. The hierarchy of sugar work, then, while notable in contrast to cotton, was still limited.

Sugar work contrasted with cotton in another important aspect: there was very little of the "tasking" associated with cotton-picking. As we have seen, some Brazos slaves negotiated a system in which picking quotas might be transferred, allowing some slaves to make up for the shortfalls of others. That practice opened a space in the work regime that was controlled by the slaves themselves even as they produced for the master. It militated in favor of solidarity among slaves and undercut the competitive tendencies that might otherwise have emerged under such a system. In contrast, other than chopping wood for fuel and perhaps a few similar chores, very little sugar work was tasked. The complexity of sugar production, involving not only cultivation and harvest but also processing under quasi-industrial conditions, made it difficult to divide it into discrete tasks.[63] The combined efforts of all workers went into every hogshead of sugar, and it was impossible to connect any single portion of the product with any single laborer.

Sugar work therefore influenced the master-slave relationship in ways far more complex than other crop regimes did. On the one hand, the hierarchy and the industrial production techniques encouraged slaves to see themselves as individual agents. Sugar-making was, furthermore, a social process in which workers were effectively, as Marx would have expressed it, "alienated" from the product of their labor. Yet these very same characteristics could just as easily have bolstered a sense of community solidarity: the occupational hierarchy was fleeting, even over the course of a single day, let alone an entire season; the shared experience of coerced labor in a crop that belonged only to the planter might counterbalance the atomistic tendencies flowing from alienated labor. The effect of sugar work on the master-slave relationship might best be seen in nondeterministic terms, as neither totally individualizing (and therefore tending to strengthen bonds between master and slave), nor wholly conducive to community. Instead, it introduced to the master-slave relationship a broad set of potentialities—or to phrase it in terms of language, a rich vocabulary of possibilities—for use in negotiations, present and future.

The subject of these negotiations was essentially the same as with cotton: community time, control over labor conditions, and material well-being. The evidence from the Perrys' Peach Point Plantation, which converted from cotton to sugar production around 1850, suggests that the concessions received under King Cotton were retained and adapted with the switch to cane. James F. Perry, as we have seen, negotiated an arrangement with his slaves that allowed them to pick cotton on their own time and also to transfer unfulfilled picking tasks to other members of the plantation community. At about the same time as the conversion to sugar, Perry and his slaves apparently reworked the agreement to compensate for the fact that cotton was no longer being produced. Instead, Perry subdivided part of his plantation into nine one-acre plots, one for each head of a slave family. He never recorded what his slaves did with the plots, but they presumably used the land to grow food crops. Also, as we have already seen, Perry paid at least some of his skilled workers a cash incentive.

Although it is possible that Perry and other Brazos masters extended such privileges in accordance with paternalistic notions of the "good master," it was almost certainly at the insistence of the slaves, with Mexico firmly in their heads, that the concessions were made.[64]

"Voodoo and Conjure Doctors"

If paternalism was the language spoken along the border between the world of the masters and the world of the slaves, it is much more difficult to say what words, metaphors, narratives, and parables structured life in the slave quarters. Interviews with former slaves, conducted in the 1930s, provide the best glimpse of what went on when the master was out of earshot, but they do not reveal everything. The fact that most of the interviews were conducted by whites with elderly black men and women in a decidedly white-supremacist environment several decades after emancipation dictates caution. In addition, the opportunities for the enslaved to practice and elaborate a distinct culture varied over the four decades between settlement and emancipation. These opportunities varied geographically as well, from the Upper to the Lower Country and from the river bottoms to the uplands. Lastly, it is important not to romanticize the slave community. The existence of the shared and distinctive values that define community does not imply perfect harmony. In fact, the opposite is often true: the closer people are, the more bitter their conflicts. Discord and personal animosity are the very stuff of community life. Nevertheless, as imperfect as our picture may be, we can reasonably say that most Brazos slaves enjoyed a rich and distinct community and cultural life, one that nurtured values that stood in opposition to those of the master class and which belied the stunted tropes of paternalism.

In the 1820s, a majority of Brazos slaves lived on small farms and plantations. In both the Upper and the Lower Country, most masters had few enough slaves so that both free and enslaved members of the household could have lived under one roof. Such cramped accommodations would have made it difficult for slaves to find the time and space necessary for a separate and distinctive community life. The small holdings and physical closeness probably represented a continuity of sorts for the large proportion of Brazos slaves who came from the Upper South, where over the previous century Anglo and African American cultures had grown close. Accounts of early Brazos community life often stress the closeness and interpenetration of white and black cultures. Noah Smithwick described a community dance in 1827 at which a slave "performed on a clevis as an accompaniment to his singing, while another negro scraped on a cotton hoe with a case knife." On another occasion, R. H. Williamson, a white man and judge who lived near San Felipe, entertained a mixed-race crowd by "patting Juba," an African American musical art in which the practitioner used his hands to slap a complex rhythmic accompaniment against his legs, chest, hands, and arms. Williamson brought the house down by incorporating his wooden leg into the routine. Dances and frolics were by no means the only occasions for interracial community activity. Revival meetings, conducted openly only after independence from Mexico in 1836, attracted both whites and blacks, though they were probably seated in separate areas. Methodist minister Joseph P. Sneed, who held a camp meeting along New Year's Creek in 1839, noted approvingly that along with thirty-nine white converts, "a good many colored and equally that number found pardon to God be all the Glory."[65]

Yet opportunities to practice and elaborate a separate, black culture existed for some, even in the earliest years. Although a majority of Brazos slaves in the 1820s lived with fewer than twenty other slaves (and most of those on holdings smaller than ten slaves), the fact that many households employed white "laborers" and "renters" often made an extra dwelling necessary. These dwellings may have furnished the space for community life apart from the master's family. For example, John P. Coles's household included five nuclear family members, five laborers or renters, and four slaves. The fourteen members of his household almost certainly moved into multiple cabins as soon as was practicable, with Cole's white laborers likely living separately from his black slaves. Moreover, the few planters who owned more than twenty slaves accounted for 42 percent of the total enslaved population. Jared Groce's Bernardo Plantation, with ninety slaves, constituted approximately 20 percent of the Austin Colony slave population in 1825. The sheer number of slaves at Groce's, along with their geographic isolation, would have permitted an autonomous slave culture to flourish. In addition, the likelihood that some of Groce's slaves had been imported into the United States from Africa only seven years earlier, in 1817, suggests a partial re-Africanization of the Bernardo community.[66]

Finally, the boundaries of the slave community were not necessarily identical with those of the farms and plantations. The sparseness of settlement along the Brazos, especially in the early years, may have prompted more inter-plantation contact, with slave owners sending trusted slaves—almost always male—off the premises on business. Groce sent his first cotton crops into the Mexican interior via mule train using enslaved teamsters. James F. Perry routinely paid the ferriage charges for his slaves when they crossed the Brazos River on various errands. The practice of slave hiring linked communities as well, with one resident even hiring slaves to serve as boatmen. Many more moved between quarters without the master's permission, either as short-term absentees or as runaways seeking permanent freedom deeper into Mexico. Slave mobility, though limited, extended community networks over large areas and allowed slaves on even the smallest farms some contact with other slaves. This was especially true of the Lower Country.[67]

As Brazos farms and plantations grew, the ability of slaves to participate in autonomous community life increased. As early as 1837, 58 percent of Brazoria County slaves lived on plantations with twenty or more slaves, and by 1850, more than 70 percent of the population was enslaved. The large plantations inhibited the growth of white-controlled public institutions, such as churches. As late as 1850, there were only four public houses of worship in the Lower Country; all four were in Fort Bend County and none in Brazoria County. A small number of individual plantations had churches, and itinerant preachers passed through from time to time, but many Lower Country planters were suspicious of them and prevented them from ministering to their slaves. Thus, the cultural and spiritual lives of Lower Country slaves were largely untouched by whites.[68]

Just as important, perhaps half of the 1830s slave population was composed of recent arrivals from the Bight of Benin or West Central Africa who were concentrated on a few plantations, several of which abutted other plantations with African-born slaves. The prominence of Africans on Lower Country plantations in the 1830s raises the possibility that the area's early slave culture was influenced very strongly by African—in this case Yoruba and Kongo—folkways (see chapter 1). However, very little evidence of such an influence survives. The paucity of evidence does not necessarily mean that the Africans had little impact or that Yoruba- and KiKongo-speakers were unable to continue their respective cultural practices. The concentration of Africans of probably similar background suggests that they did to some extent, especially at the level of the household, but it is difficult to disentangle specific ethnic practices from more general African and African American practices.[69]

A small body of archaeological evidence has raised the possibility of specific cultural continuities with Africa, in this case, West Central Africa. Slave-quarter excavations at the Levi Jordan sugar plantation in Brazoria County by archaeologist Kenneth Brown have uncovered in the subfloor of one of the dwellings coins that appear to have been deliberately placed to form a quadrilateral. In addition to the coins, Brown excavated several knife points and small porcelain dolls, which he suggests were similar to those used in divination ceremonies in West Central Africa, as well as two cast-iron kettles stacked together. Brown argues that these artifacts "support the interpretation of an African and African American behavioral/belief system." Specifically, he suggests that the coins were placed to form a four-point cosmogram, a graphic representation of the Kongo spiritual universe, and that the knife points, kettles, and figurines "appear to be all that remains in the archaeological record of a wooden Nkisi," or shrine to a Central African local spirit.[70]

Given the probable regional origins of the Africans in Brazoria, Brown's interpretations have some plausibility. It is very likely that a sizable proportion of the Africans in Brazoria County originated in West Central Africa, and therefore it is possible that the artifacts at the Jordan plantation actually do demonstrate the transplantation of West Central African spiritual understandings to the banks of the Brazos. But there are other possibilities that undermine that conclusion. First, although the four-point cosmogram held a special significance in Kongo culture, other West African traditions employed similar iconography. Specifically, many Yoruba-speakers placed a special emphasis on the notion of a crossroads, the place where the Yoruba gods had once ascended to the heavens. One of the most important orishas, or deities, in the Yoruba sacred tradition was Eshu-Elegba, the god of the crossroads, who controlled access to the other orishas. Yorubas also used iron kettles and cauldrons like those found at Jordan's in their ceremonies in association with Ogun, the god of iron and war. Ogun was also associated with the clearing of land and the use of the machete, activities that consumed a great deal of Lower Country slaves' energies. Whether the coins represent a cosmogram or a crossroads, or whether the kettles were part of an *nkisi* or were used in a ceremony involving Ogun, or whether they represented a creolized fusion of Kongo, Yoruba, and African American spirituality is impossible to say. In short, the objects at Levi Jordan's—if indeed they were placed by Africans—tell us little about Africans' origins, let alone how they made sense of the cosmos.[71]

Adding to the general uncertainty about the significance of the artifacts excavated at Jordan's is the lack of any evidence that African-born slaves actually lived on his plantation. Although it is possible to document the presence of African-born slaves on many of the nearby plantations, including a dozen or more Africans at the adjacent Fannin-Mims place, no contemporary evidence places Africans at Jordan's in the antebellum period. Levi Jordan arrived from Arkansas with a full complement of slaves in 1848, approximately twelve years after most of the Africans on Brazos plantations arrived. It is possible that Jordan purchased or hired a small number of African slaves after arriving, though probably not many, and it is equally possible that African-born slaves from the Fannin-Mims plantations spent some time at Jordan's. But given the circumstances, it is more likely that the objects at Jordan's represent the creolized African heritage of the Jordan slaves, possibly in combination with the Old World expressions of African-born neighbors.

In contrast to the archaeological record, the ex-slave interviews conducted during the 1930s make it clear that African-derived spirituality flourished on the Brazos, but they are similarly mute on their specific ethnic origins. Lucy Lewis, who lived on several of the McNeel plantations, where an undetermined but significant number of Africans also resided, told her interviewer that some of the slaves "wore charms round dey necks, little bags of asfeddity." While certainly of African origin, such practices are impossible to connect with specific African peoples. Charms and amulets in the form of small bags were employed throughout West Africa, and even Muslims used them to hold verses from the Koran. Whether the charms, along with the many references to "voodoo" in the WPA narratives, speak to a particular African or creole religion or to a generalized image of African-derived divination and spiritual practice is impossible to say. The interview with Patsey Moses, born in 1863 in Fort Bend County, reveals just how knotty the issue could be. Although she told her interviewer that "voodoo and conjur doctors" had the "mos' power" in the black community, there is no evidence that she ever lived with African-born men and women. One of the conjurers, she recounted, "had a voodoo kettle" and held ceremonies "on the square" at night, a possible reference to the four-point cosmogram or the crossroads. But Moses stated very clearly that she had never personally observed these particular goings-on—in fact, they had occurred in Knoxville, Tennessee, and she had only learned of them after the fact. So although the sources hint at the possibility of African ethnoreligious transplantations, they do not allow us to disaggregate African and African American practices.[72]

If the effort to document specific African transplantations along the Brazos seems doomed to frustration, it is certain that a more general concept of Africa loomed large in the consciousness of all Brazoria slaves. Three of the seven former Brazoria County slaves interviewed by the WPA actually mentioned Africa or Africans. The portrait that emerges is sketchy, what we would expect a full century after the Africans' arrival. Cinto Lewis (Lucy Lewis's husband), who had lived on David Randon's plantation, stressed the alienation of the Africans from the creoles, a theme already suggested by the Africans' marital endogamy. "Some niggers jus' from Africa," his interviewer quoted him as saying, "and old Marse has to watch 'em close, 'cause they is de what mostly runs away to de woods."[73] Another interviewee, Nancy Antwine, the daughter of African-born parents, suggested a more complex but ultimately more harmonious relationship between the two groups. In her interview she told the story of her mother's enslavement (though not her father's). According to Antwine, her mother and her mother's sister had gone down to the river to wash peanuts. A boat suddenly appeared and the "white folks" on it asked the girls if they wanted to trade the nuts for "lots of red cloth and things." They boarded the boat, but it left before they could debark. The boat eventually stopped in a "big place," where Antwine's mother could see "lots of more black folks."[74] At that point the two sisters were separated, never to see each other again. Antwine's story was apparently repeated so many times over the years that it resembled a parable, with its moral and symbolic qualities overshadowing the specific personal histories. The most obvious function of the tale was to keep the memory of Africa alive by representing the true experience of a member of the local slave community. More importantly, the tale offered a coded critique of the slave regime and its origins. Significantly, Antwine did not use condemnatory language in her retelling. Noting only that that her mother's captors were white (which was unlikely to have been the case), Antwine exhibited no ill will toward them. She did not say that her mother was "kidnapped" or "captured;" rather, she "came from Africa." Similarly, the boat simply "starts goin'" while Antwine's mother and aunt were "lookin' around." She did not mention violence and did not give details of the Middle Passage. In fact, Antwine's story comes sandwiched between two paragraphs that stress how good her "white folks" were. The white interviewer found the story "interesting," but the tale's moral relevance should have been clear to anyone who bothered to listen closely. First, it offered a historical explanation for slavery, a Book of Genesis that told what happened In the Beginning. Second, while Antwine's recollection that the captors were white probably did not reflect what actually happened, it did articulate a certain moral principle: as the ultimate beneficiaries of black labor, whites assumed the ultimate culpability for slavery. In other words, Antwine's narrative of events, if not her choice of words, condemned slavery. Her ability to tell whites what they needed to hear, honed over her long life, was still sharp. Lastly, Antwine's statement that her mother arrived at a place with "lots of black folks" probably speaks to the eventual solidarity that emerged between African-born slaves and creoles, an identity rooted primarily in race rather than ethnicity.[75]

What, then, does the evidence say about Lower Country slave community and culture? First, demographic conditions were conducive to the formation of a separate and distinct slave culture. Plantations were large, with a median population of fifty, and seven out of ten residents were enslaved. Second, while Lower Country slave culture was almost certainly Africanized to a degree greater than virtually anywhere else in the United States, the evidence, both archaeological and historical, is too sketchy to identify specific African practices with any confidence. Evidence that Africans practiced marital endogamy suggests that distinct ethnic practices may have continued in specific households, but whether they continued at the community level is much less certain. Third, some notion of Africa occupied a central place in Lower Country slave culture, but its significance changed over time. Early on, as suggested by Cinto Lewis's narrative, Africans and creoles found relatively little common ground. Later, as suggested by Nancy Antwine's narrative, as the Africans became more familiar with the English language and the ways of southern plantation life, and as they began to have creole children, the notion of a shared African past and shared blackness probably helped bridge earlier gaps. Finally, the existence and strength of the Lower Country slave culture was oppositional, allowing slaves to articulate a critique of slavery and to anticipate a future without slavery, whether in Mexico or at home.

Upper Country slaves faced a different set of challenges and opportunities. Slaveholdings remained small in comparison with those of the Lower Country until the growth spurt of the 1850s. Up to that time, most slaves would have lived and worked in close contact with their masters, their opportunities to enjoy time with other African Americans limited to illicit visits and the occasional sanctioned trip off the premises. After 1850, however, Upper Country slave quarters expanded significantly. Many slaves also resided apart from their masters. Several of the Chapell Hill planters, for example, chose to live in or near the town while many of their slaves lived on noncontiguous bottomland properties a few miles away. Planters in the Navasota area of Grimes County seem to have done the same thing. Thus, in neighborhoods that already had a majority of African Americans, the black population was further concentrated on quarters that were often several hours' ride from the master's residence.

Several factors distinguished Upper Country cultural and spiritual life. In contrast to the essentially unchurched landscape of Lower Country, the Upper Country featured more public houses of worship, providing more opportunities for whites and blacks to share a religious culture. Even so, with only one church for every twenty-seven hundred residents, a minority of slaves (and whites) would have been able to worship regularly in public together. Evidence from the WPA narratives reflects this situation. Lizzie Atkins of Grimes County recalled only annual camp meetings, after which "those converted could be baptized, then we went home happy both black and white until the next year." Steve Robertson of Washington County told his interviewer that as an apprentice coachman he went to church regularly, but the remainder of the slaves on the plantation never attended. Black religious life, and community life in general, was thus largely its own independent domain, with only marginally more interference from whites than in the Lower Country.[76]

A second distinguishing factor for the Upper Country was the nearly total absence of African-born slaves, making the local slave culture significantly less African and more creole in nature. Although some interviewees from the 1930s referred to charms and conjuring, the proportion is notably smaller than for the Lower Country. In the Upper Country, conjuring and charms complemented a mostly Christian worldview, though one very different from that of the masters. As happened so often, master-sanctioned sermons rang hollow in the ears of slaves. Louvinia Young Pleasant, who lived in Chappell Hill, recalled that her master dispatched a preacher to the quarter about once a month. "About de only thing he would preach would be: 'Now, don't yo' all steal yo' mawster's eggs. . . . Now don't yo' all steal yo' mawster's sugar. . . . Now don't yo' all steal yo' mawster's chickens,'" she recalled, with palpable contempt.[77]

According to the southern ideal, the masters dictated the tone and content of the master-slave relationship. Reality, of course, was never so neat. Slaves may not have gotten all they wanted from their masters, but they were able to influence the day-to-day workings of the institution and practice an independent and distinctive culture of their own. However, in addition to the daily give-and-take between masters and slaves, there was a third shaping influence that lay beyond the boundaries of the plantation—other Brazos residents, some of whom owned slaves and some of whom did not. It is to those relationships that we now turn.

Chapter 4

GERMANS, ANGLOS, AND THE POLITICS OF SLAVERY

The Brazos was unusual among antebellum plantation regions in its ability to attract a significant number of immigrants. Several factors were responsible: liberal Mexican land policies in the 1830s, coupled with a preference for non-U.S. immigrants; the founding of several immigrant enclaves just a few years before plantation agriculture took hold; and a sustained chain migration from several parts of German-speaking Europe. In the 1830s and 1840s, Germans and Anglos enjoyed good relations. This fact was actually quite significant in an ideological sense. Rural southern culture placed a premium on neighborliness, and to a great extent Anglos imagined the political process to be a way of affirming and regulating neighborly relations. But it was slavery that made Anglo southerners particularly sensitive to their neighbors. In order to function, slavery required not just an accommodation between master and bondsman, but an agreement that all free people would respect the master-slave relationship. Slaveholders needed to recognize each other's authority and refrain from interfering with each other's laborers, and nonslaveholders would have to do the same. In general, they did: throughout the South a tacit understanding existed among slaveholders and yeomen that the head of a household ruled it as his own domain. No interference with household dependents, be they slaves, wives, or children, would be tolerated.

Of course, the ideal was one thing and practice often another. In reality, people tampered with others' slaves on a regular basis. In normal times, slaveholders regarded interference as a nuisance, or at most an affront to their honor, but not as a threat to the institution of slavery itself. Yet as the sectional crisis unfolded, and as the German population increased, Brazos slaveholders became obsessed with their neighbors' views on the question. Germans, while never condemning slavery outright, generally did not embrace it and showed little support for the sort of Southern Rights politicians who gained popularity in the 1850s. Upset at German unsoundness on the slavery question, Anglo residents began to worry that their neighbors were in fact a cadre of crypto-abolitionists. For Anglo slaveholders and their allies, the immigrant presence made the abolitionist threat a very real one. Ironically, the more the Germans came under attack for their alleged softness on the question of slavery, the more they were driven into real political opposition. Neighborly reciprocities, German-Anglo relations, and sectional politics thus fused into one, adding fuel to secessionist fires.

Neighborly Politics

Brazos residents, like many other rural people, idealized what might be termed a code of neighborliness. Neighborliness was grounded in the assumption that heads of households would respect each other's prerogatives: good neighbors did not interfere with each other's dependents; intervention was tolerated only under the most unusual of circumstances. Neighborliness also implied a large, but vague, set of interhousehold reciprocities. Most often it was invoked in the breech, such as when one neighbor's livestock created a nuisance on another's land, a frequent occurrence in open-range societies like the Brazos. A romp by a young colt through the fields of Chappell Hill's William W. Browning was the occasion of a typical neighborly exchange. When Browning complained, the owner, B. D. Dashiell, removed the animal immediately, telling Browning, "I feel under many obligations for the priviledge granted me and regret that the colt have caused you any trouble."[1]

Rebecca Pilsbury, a native of Massachusetts who lived on Gulf Prairie, learned of her obligations on the occasion of a neighbor's illness. The neighbor, a Mrs. Blackwell, was upset that Rebecca's husband, U.S. Representative Timothy Pilsbury, had left for Washington, DC, without calling on her. Rebecca visited Mrs. Blackwell the day after her husband departed and found her "quite grieved" over the matter. "I fancied she attributed his failing to do so to me," Rebecca wrote in her diary, "and if I am not mistaken she queried whether I would be *neighborly* during his absence." Eager to please Mrs. Blackwell, Rebecca visited her regularly over the next few months. Just before Christmas, she even sent several slaves to Mrs. Blackwell to help her kill some hogs. "Quite a tax I think," wrote Rebecca, "but I have no doubt [Mrs. Blackwell] thinks it an honorable and honored custom." That Rebecca Pilsbury's overture involved sending "the boys" to Mrs. Blackwell underscores another important feature of the neighborly economy: the medium of exchange was often enslaved labor. To be certain, nonslaveholding households exchanged labor, sending members to gather and transport crops and raise out-buildings, but the lending of slaves was a particularly neighborly thing to do because they could be worked harder and at particularly nasty tasks like hog-butchering, all without incurring any hard feelings or worry that one neighbor was taking advantage of the other.[2]

The ideal of neighborly consensus informed the early political culture of the Brazos. In the political imagination of the Anglo South, household independence was the bedrock of republican liberty. Political sovereignty rested, in turn, on the notion that the head of household must be left free to exercise sovereignty over his dependents—wives, children, slaves, and servants. But the system was viable only as long as heads of household agreed, in politics no less than in neighborly encounters, that the various domestic institutions were legitimate. It would simply not do for one neighbor not to recognize another's claim to his wife, his children, or his slaves. As a result, Anglo southerners placed a premium on neighborhood consensus and looked askance at heterodoxy. Put another way, there was a normative expectation of harmony, even if the normal state of affairs was not always harmonious. Mary Austin Holley captured the connection between neighborliness, politics, and the desire for harmony when she wrote in 1831, "All [in Texas] are happy because busy; and none meddle with the affairs of their neighbors, because they have enough to do to take care of their own. They are bound together by a common interest, by sameness of purpose, and hopes. As far as I could learn, they have no envyings, no jealousies, no bickerings, through politics or fanaticism."[3]

Political resentments thus focused externally on Mexican political control, which grew increasingly intolerable to most Anglos precisely because it appeared to threaten household sovereignty. Concern that the national government would abolish slavery once and for all vexed and traumatized slaveholders. But Anglos discerned serious if less direct threats to the integrity of their households as well, specifically in General Santa Anna's strike, starting in 1835, at the federalist provisions of the constitution of 1824. Although at first glance there might seem to be little connection between Mexican centralization and household sovereignty, Anglos, particularly those from the southern states, would have seen one. During the Nullification Crisis only two years earlier, many white southerners drew explicit connections between a powerful U.S. national government and the cause of antislavery. Nullification failed, but southern politicians waged a largely successful fight to ensure that the federal government could never be used by outsiders to undermine the authority of a master over his slaves, instituting, among other things, a gag rule prohibiting the reading of antislavery petitions in Congress and prohibiting the distribution of antislavery literature through the federal mails. The impetus for Mexican centralization grew out of completely different historical circumstances, but Anglo residents tended to view it with their own recent struggles in mind, fearing interference from afar.[4]

Where political conflict existed, it tended to manifest itself in locally based personal factions rather than in well-organized political parties, and as a rule it never threatened the fundamental consensus on household sovereignty. During the 1830s, as Brazos residents debated their response to Santa Anna's centralization of political power, two factions, known as the War Party and the Peace Party, squared off against each other. Their rhetoric was heated. Gulf Prairie planter John G. McNeel considered the war faction to be composed of "aspirants and speculators, [who] are attempting to deceive the people." War hawks tarred their opponents as "Gentlemen" who secretly hoped to institute a property requirement for voting. Although the conflict might appear to be class based, the contestants sorted out largely according to their relationship with Stephen F. Austin: those whose prosperity owed largely to their ties with the empresario, including virtually all of the Gulf Prairie planters, gravitated to the Peace Party; the others, a sizable majority that included many recent arrivals, supported political independence. In fact, the most outstanding characteristic of the time was not the conflict between the War Party and the Peace Party, but the overwhelming agreement on maintaining the order and independence of the region's households. When a Mexican force under General Martín Cos came to forestall independence, the factions melted in the face of the common threat, and an independence movement was born.[5]

Politics during the years of the Texas Republic (1836–1845) played out similarly in their preoccupation with the connection between household and political independence. The key political question of the day, the relationship between Texas and Mexico, hinged partly on the issue of whether the Mexican government would accept the loss of its northern province. The political independence that was guaranteed by locating the international boundary at the Rio Grande articulated with deeply held notions of household mastery. In gendered language redolent with Jacksonian imagery, pro-independence speakers and editors in the 1830s had urged Texans to refuse to "yield in servile submission to a form of government or to anything else that would disgrace us as free born men."[6] The appeal was all the more attractive to those who had been in Texas before 1836 and recalled the threat that Mexico had posed to slavery. For most Anglos in the era of the Republic, annexation by the United States was desirable because its republican federal system was a hedge against precisely these abuses. The alternative, a British guarantee of independence in exchange for an emancipation, would have represented for many a violation of the most fundamental principles of household mastery and political liberty. Emancipation would have deprived the owner of his lawful property and the master of authority over his household. With political independence for Texas underwritten by Britain, liberty would be a sham. Brazoria's representative, Guy M. Bryan, very tellingly called the plan a "humbug of [the] name of Independence"; he referred of course to its political shortcomings but was undoubtedly aware of its domestic implications as well.[7] With annexation in 1845 and the Treaty of Guadalupe Hidalgo in 1848, Mexico ceased to be an important political issue along the Brazos, frequent grumbling about the harboring of fugitive slaves notwithstanding.

The political questions of the 1830s and 1840s were debated not by parties, but by personal factions. The two most enduring of these, led by Sam Houston and by Mirabeau B. Lamar, bore a crude resemblance to the Democrats and Whigs in the United States, respectively, although not without important differences. On Indian policy, for example, it was the Jackson protégé Houston who was the more accommodating, while Lamar pursued a much more aggressive course. In most other respects however, Houston men mirrored their Jacksonian counterparts, favoring paying off the public debt and strict limits on the granting of corporate charters, while Lamar men agitated for a central bank. But the Houston and Lamar factions were never well organized. Elections often featured multiple candidates, some running as local or regional favorites. Canvasses were very much neighborhood affairs, characterized by barbecues at which office-seekers were expected to "treat" gatherings of voters to copious amounts of whiskey. Fistfights were a standard feature, as were various other contests of strength and chance. Candidates who did not treat the crowd could expect to be branded by opponents, as Guy M. Bryan was in 1847, as "too aristocratic to treat and drink with common people." Bryan nevertheless won that particular election, testifying to the power of personal faction and the politics of neighborhood in mitigating whatever class tensions existed among Anglos.[8]

The neighborliness of the 1820s and 1830s, though idealized, came under considerable strain after 1845, and especially so during the 1850s, because of the spread of market-oriented agriculture. Nineteenth-century Americans were no strangers to the market. For many, however, the move to Texas represented an effort to regain a measure of control of household affairs that had eroded considerably in the decades after 1815. In effect, a move through space to the southwest was also a bid to travel through time, back to an era in which a man mastered the market and not the reverse. Observers in the early 1830s still commented on the unfailing generosity of Brazos residents, noting for example that an offer to pay a farmer for food or lodging was universally taken as an insult.[9] By the end of the decade, however, some of the "Old Settlers" were beginning to complain about the "new race [of] adventurers, sharpers, and . . . blacklegs" that had come to dominate.[10]

The final antebellum decade saw a remarkable expansion in staples agriculture along the Brazos, in gross quantities as well as in proportion to food production. The most striking changes occurred in the two largest counties in the Upper Country, Washington and Grimes (see table 9). Between 1849 and 1859, Grimes County farmers and planters increased their production of cotton in relation to corn by a factor of three. Washington County farmers' ratio of cotton to corn almost doubled in the same ten-year period, and in certain neighborhoods, such as Chappell Hill, the ratio increased faster. Even in the Lower Country, where more reliable river transportation had allowed plantation agriculture to take root a decade or more earlier, a steeper tilt toward staples production was noticeable.

Many residents experienced the spread of market relations and of staple-crop concentration and the decline of neighborliness as a moral crisis. The feeling was most acute in the Upper Country counties of Grimes and Washington, where the rise of staples-oriented agriculture was the most sudden. For evangelicals, who drew a sharp distinction between the sinful World and the Kingdom of Christ, it appeared that selfishness had overcome charity. In an article titled "The World Is Hostile as Ever," a writer for the *Texas Baptist* in Anderson lamented that "Business is now carried on with a zest and intensity which were unknown in former days; time and thought are so completely absorbed by it, that neither leisure nor heart are left for anything higher; a spirit of cold selfishness is thereby engendered, even among those who are called christian tradesmen."[11] In addition, staples agriculture fostered what appeared to many to be decadent and worldly behavior. In order to grow cotton, farmers and planters bought tools, seed, and slaves on credit, which stirred condemnation among evangelicals. "When cotton commands a good price," lectured the *Baptist*, "the production of provisions is more and more neglected—debts are incurred—land, negroes, mules, and almost everything rises, and if nothing else operates to effect [*sic*] the price of cotton, the expansion cannot last long, and ere long a contraction commences." The solution was simple: for a planter "to be sure always to raise provisions to use, and some to spare, and then make as much cotton as he can to do justice to himself and the country."[12]

The prodigious growth of the 1850s also strained neighborly relations by widening the distance between planters and plain folk. The gap was simultaneously social and spatial, as certain neighborhoods, especially in the bottoms, developed reputations as planter areas while others were inhabited primarily by yeomen. John Washington Lockhart's primary social circle in Chappell Hill, for example, consisted of his hunting partner Frank Toland and Jonathan Wallis and William Traynham, with whom he and his family frequently dined. All were quite well off and owned 150 slaves among them.[13] In addition, many elite families found themselves united by marriage. Lockhart's family alone had ties to several elite families in Chappell Hill, including the Brownings, the Wallises, and the Tolands. Parties and balls were also generally limited to a select set of wealthy families. Oscar Addison noted the exclusive character of an evening tea hosted by Thomas Masterson of Brazoria County. "The first person to arrive was a common man from town, & I was surprised that he should have been invited," he wrote, but when Addison realized the man was a hired musician, his surprise was dispelled.[14]

Public labor arrangements placed strain on neighborly relations as well. Like most southern states, Texas had no state agency responsible for building and maintaining roads and other public works. Counties were divided into road districts, each under the supervision of an appointed "overseer." The overseer's job was to make certain that each district resident contributed to the upkeep of the roads by contributing labor or laborers. A road duty list from the 1850s makes it very clear that slave and free laborers worked together. "Should there be any hands on this road subject to road duty, not on the above list, you must order them out to work," read one set of instructions, "White or Black."[15] The insult of the system, which essentially required white nonslaveholders to do road duty while planters simply detailed their bondsmen, could not have been lost on Brazos plain folk. Nineteenth-century southern plantation society required the cooperation of white nonslaveholders, which they gladly gave as long as they felt their own independence and equality were secure. Inequality, however, lay at the heart of plantation society, and containing it proved difficult at times.

Anglos dealt with the uncertainty of the 1850s by attempting to recreate neighborly relations through more formal means. The founding of several agricultural associations was emblematic of this impulse. In general, the associations provided a forum for the exchange of information on the best cultivation methods, brands of seed, and breeds of stock; and established "premiums," or prizes, to be awarded for the best produce. Some organizations, such as the Brazoria County Agricultural Society, gave voice to the growing perception of conflict between planters and merchants. Members exhibited little embarrassment in pursuing their particular interests, declaring that the object of the society was simply to "protect and advance all the interests dependant or connected with those of the planter." For the most part, however, formal agricultural societies tended to harmonize conflict by focusing on common pests, by testing different agricultural methods, and by stretching the definition of "planter" to include even the most modest farmers.[16]

Evangelical churches were another medium for mitigating potential conflict between planters and plain folk. Because Protestantism was not tolerated before 1836, few churches existed along the Brazos. As late as the 1840s, many neighborhoods were still dependent on occasional visits by itinerant preachers, though when one did pass through it was a major event. An 1839 camp meeting held by Methodist Joseph Sneed near the eventual site of Chappell Hill attracted worshippers from more than a hundred miles away. Several neighborhoods actually began life as church congregations. Chappell Hill was one of these, as we have seen, centered initially on Providence Baptist Church. Interestingly, during the 1850s Chappell Hill shed its Baptist identity but not its evangelical character, and by 1860 it was known as a regional center for

Methodism, with members encompassing the wealthiest planters as well as property-less clerks and laborers. Church membership in Chappell Hill had grown, but not merely because the population had increased. Evangelicals stressed brotherhood in Christ, which bore a strong resemblance to the secular code of neighborliness. During this period of expanding commerce and growing inequality, the appeal must have been powerful indeed. The connections between market-oriented staples production, agricultural societies, and churches are suggested in an article in the constitution of the Grimes County Agricultural Society granting membership to "all evangelical ministers of the gospel of this country." It was through mechanisms such as these that Anglo whites attempted to maintain the neighborly ideal.[17]

Interference

Although white Brazos residents considered slaves to be members of their households and therefore not neighbors in their own right, slaves frequently formed relationships that extended beyond the boundaries of the households to which they belonged. Most slave owners hoped to minimize their slaves' outside contacts, but forbidding them altogether would have eliminated an important currency in the perpetual negotiations between master and slave. Abroad marriages were among the more common forms of interhousehold relations among slaves, and among the more visible because they were often tolerated by masters. Slave owners also occasionally allowed their slaves to attend frolics at neighboring farms and plantations, sometimes even to attend special events in nearby towns. One such event, a dance held on the outskirts of Washington, attracted a sizable number of slaves from the surrounding area. The dance was interrupted by a local white man who grew jealous when his mulatta "paramour" danced with a black man. The intruder smashed the orchestra's violins and departed, prompting the dancers to complain to their masters. A group of "respectable citizens" soon rallied to protect the slaves' right to enjoy "authorized amusement" and set out after the crasher. Some time later, the man returned and stabbed one of the slaves' white "protectors" to death, then made a second escape. The good citizens of Washington petitioned Governor Peter H. Bell to take steps to apprehend the killer, emphasizing that the event was sanctioned by local masters and that all attendees had passes.[18]

Far more troubling from the slave owners' point of view was the tendency of unfree Brazos residents to forge unauthorized bonds with their free neighbors. The stability of the slave system demanded the unwavering complicity of nonslaveholders, although throughout southern history there was never a shortage of whites willing to undermine the exclusivity of the master-slave relationship by trading in pilfered plantation goods, engaging in sexual relations (both consensual and not), and offering aid and comfort to fugitives. This sort of activity drove slave owners to distraction, and southern legislatures churned out a steady stream of laws designed to prevent it. But nothing could stop one of the most common instances of extrahousehold contact, which occurred when slaves prevailed on neighboring whites to intervene on their behalf. Situations in which the master's authority was divided, such as slave hiring, were especially conducive to manipulation by slaves.

Gideon Lincecum of Washington County learned this lesson when a slave of his named Dan got into a dispute with some slaves belonging to Lincecum's neighbor and son-in-law, Matson. Matson sent a note to Lincecum telling him that he was considering having Dan prosecuted and suggesting that Lincecum was shielding Dan from the authorities, a charge Lincecum denied. Lincecum pointed out that in a previous instance in which one of Matson's slaves attempted to poison one of Lincecum's, he had declined to prosecute. "But to relieve your mind on the subject," offered Lincecum, "I hereby give you clear liberty . . . that if ever you find that the said Dan on any part or portion of your premises with a pass or without a pass . . . to shoot him, or hang him, or do whatever else you may think proper with him." In other words, Lincecum insisted on the inviolability of the household—masters were to control and punish their own slaves without interference, unless they trespassed on another's domain, in which case the "liberty" to defend one's dependents took precedence. Lincecum ended his lecture to his son-in-law by excoriating him for letting the neighbors know of their dispute. "This is all private," he said, "but if you desire it and cannot be satisfied short of notifying community of these little family matters, go ahead, and we may succeed in procuring for ourselves and those who are connected with us, quite an unenviable notoriety."[19]

However, contrary to Lincecum's claims, in certain instances interference with another's slaves was condoned, usually with an eye toward strengthening the system as a whole. In criminal matters, for example, slaves could be placed on trial, and in capital cases an 1852 law compensated the master with up to one thousand dollars as a way to discourage the shielding of defendants. The legislature declined to authorize compensation for any slave killed in mob action, though—to do so might encourage citizens to "take the law into their own hands" and ultimately weaken, rather than strengthen, the master-slave relationship. And, as in other slave states, Texas law provided for local slave patrols, empowering members to take up any white person found engaging in "amusements or associations" with slaves and to mete out up to twenty-five lashes to any slave found abroad without a pass.[20]

The German Element

These sorts of incidents occurred frequently across the South without ever prompting serious concerns for the viability of the peculiar institution. But the presence of so many Germans, whose soundness on slavery was never quite certain, made such contacts more fraught along the Brazos than elsewhere. The steady flow of German immigrants, which accelerated dramatically in Austin and then Washington County in the 1850s, made neighborly relations on the Brazos extremely complex. In the 1830s and 1840s, when their numbers were small, German immigrants passed relatively easily into the matrix of neighborhood. With few countrymen in the area, immigrants had little choice but to associate with Anglos. Some new arrivals harbored romantic images of American "backwoodsmen" and were thrilled to encounter the genuine article. Germans admired the backwoodsmen's alleged sense of community and hospitality, perhaps because the image represented the antithesis of the commercial-individual ethos afoot in German society. Rosa Kleberg's admiration for the "American pioneers" prompted her to attend several camp meetings, as well as a dance and barbecue at San Felipe in celebration of the Battle of San Jacinto. "Reels and squares were the favorite dances," she recalled, "and I was much impressed with the loud prompting, which is not customary in Germany."[21]

German-Anglo mingling never vanished entirely, even as the immigrant community expanded in the 1850s. The hiring of German wage labors on the plantations was one of the more common catalysts for contact. One year John Washington Lockhart employed Frederich Bay, almost certainly a German, to supervise his Chappell Hill plantation.[22] Josef Silar, a Bohemian, did some overseeing as well, perhaps because he felt an outsider in heavily German Cat Spring. "Looking for a way to make a living in the world, I entered into a closer relationship with the Americans," he explained years later.[23] In fact, those who did cross the Anglo-German boundary were often outsiders in their own communities. In 1856, for example, the only Anglo member of the Agricultural Society of Cat Spring was a Yankee from Maine, perhaps the same one encountered by Frederick Law Olmsted in 1854, who criticized the local planters for their obsession with "more nigger help" and who employed two Germans and two Englishmen on his own farm.[24] Then there were Kentuckians Sam and Lem Swearingen, who had lived near Cat Spring since before it was an immigrant neighborhood and who later publicly opposed secession. They sent their children to E. G. Mätze's German school, where several of them learned to speak German.[25] Conversely, Germans living outside the immigrant communities could be surprisingly open to Anglo culture. Henry Brandt from Schleswig-Holstein was a carpenter who settled in Chappell Hill, married a woman from Mississippi, joined the Methodist Church of the South, and volunteered for Waul's Texas Legion in the Confederate Army at the outbreak of the Civil War.[26] Charles Giesecke, a merchant from Hannover living in Brazoria, married an Anglo woman, hired slave laborers, and eventually purchased one.[27]

Most immigrants, however, lived in separate settlements a few miles from the Brazos, inscribing on the Upper Country an internal boundary that grew more distinctive as time passed. Travelers who crossed this boundary noted the differences, and many commentators, especially those who were not from the South, preferred what they saw in the uplands. Housing in the bottoms, with the exception of the increasing number of plantation homes, consisted primarily of log buildings. A U.S. Army surgeon described Brazos slave quarters shortly after the Civil War as log cabins "without windows, seldom floored, and with poor roofs, forming a one-roomed cabin from 16 to 22 feet square, and this frequently over-crowded. Whitewashing and scrubbing are a novelty almost unknown, and the bedding and clothing indifferent."[28] His view was undoubtedly tainted by condescension toward his vanquished enemies, but it conforms generally to most other descriptions. The German communities projected an image of domestic comfort, combining elements of the middle-class *Biedermeier* aesthetic with *Wildniss* romance. Charles Nagel described the Kloss home in Millheim as "nestled among large trees. The abundance of flowers marked the foreigner, as his barn and his thriving field of grain and corn did the man of honest toil and fair training. . . . Even the entrance to his barn was impressive. It had an arch over the gate."[29]

In many important respects, life in the German areas really was different from life in the Anglo-dominated bottoms. German farmers, for example, did not plunge as heavily into the market as their Anglo counterparts. However, this is not to say that Germans ignored the market. The earliest settlers grew tobacco to raise cash, then moved steadily toward cotton in the 1850s. Between 1849 and 1859, farmers in Cat Spring went from producing sixty-eight bushels of corn per bale of cotton to producing only thirty-two. But even that ratio was significantly more balanced than for the rest of the Brazos (see table 9). German farmers, then, succeeded at maintaining a balance between food and cotton production, which only strengthened the self-contained and separate nature of the immigrant communities. As Charles Nagel wrote with more than a hint of nostalgia, the antebellum period was a time "before the farmer reversed the order of things to turn his place into a business for profit and from that profit to draw enough to feed and clothe his family. The family lived on and upon the farm—the surplus might yield a profit."[30]

The large numbers of immigrants and their relative insulation from markets allowed Germans to foster a strong sense of ethnic identity. Although German-speaking immigrants came from all over central Europe, regional distinctions were of secondary importance to being German, which is not surprising considering that the rise of German nationalism corresponded roughly with the emigration to the United States.[31] Schoolchildren on the Brazos were taught in High German, while a variety of dialects were spoken in daily life, ranging from southern dialects, to northern Plattdeutsch, to Swiss German. Regionalism was generally regarded as an interesting curiosity, not as a serious division. Charles Nagel, whose roots were in Brandenburg, Prussia, recalled with interest the origins of the neighboring families in Millheim—Rostock, Bonn, the Rhine, Switzerland, southern Germany—but tellingly termed the town a "German settlement."[32] Nor did religious differences divide the population. With few churches in the immigrant districts before the Civil War, a more private piety seems to have been the rule. For those who craved religious community, the only alternative was often to join Anglo denominations. Karl Urbantke was among those who did, later explaining that he had joined the Methodists not out of an attraction to Wesleyan doctrine but because there was no Lutheran church in Cat Spring.[33]

An important exception to the rule were the Czech-speakers from Bohemia, whom Millheim resident Charles Nagel described as "essentially apart" from the German community, even as they dwelled side-by-side with Germans in Upper Country towns like Millheim, Cat Spring, and New Ulm. In the antebellum period, E. G. Mätze's school in Cat Spring served Germans only, no Czechs. Many of the Bohemians arrived somewhat later in the 1850s with few resources and supplied, in Nagel's words, "the farm labor and the maids" to the more affluent German households.[34] In 1857 Nagel's family hired a "Bohemian girl," whom he described as "a fiend for work" and who did all the cooking while his mother handled the rest of the domestic chores.[35] The girl's fifteen-year-old brother, Jan Ustyník, labored for a German at an annual wage of thirty-six dollars. The employer was a tenant himself and worked Ustyník hard, never paying him and frequently leaving him hungry. After six months, on a day when the German and his wife went visiting, Ustyník ran away and returned to his parents. Sometime later he attempted to recover back pay but had to settle for an old brass watch.[36]

In spite of the occasional exploitation, overt hostility between Germans and Bohemians seems to have been rare. For many Czech-speakers, labor and service for neighboring Germans were a temporary measure that ceased once a family managed to purchase land. In this regard, Bohemians and Germans were very much alike, often pooling the resources of extended kin or even friends to purchase land in common. Eventually these plots were divided among the families. By 1860, twenty-nine of the thirty-seven, or 78 percent, of Bohemian families in western Austin County owned their own farms. Bohemian Josef Lesikar even recounted how he and a "Czech-German fellow" purchased a farm together in New Ulm in 1854.[37] Furthermore, the Bohemians were never quite as isolated as Nagel suggested. Many arrived with the ability to speak and read German, and Bohemian schoolchildren learned German in addition to English. Czech orphans were sometimes adopted into German families, and over time intermarriage helped further to integrate the immigrant population—even Charles Nagel had a Bohemian aunt. So although they initially formed distinctive communities, the passage of time saw a gradual merging of Czechs and Germans. In Anglo and African American eyes, of course, the two groups were scarcely different at all. They were "the German element," or "the Dutch."

In many respects, German community life resembled that of Anglos. Borrowing and informal exchanges of goods and labor were as important among Brazos immigrants as they were elsewhere in the region. As in other parts of the South, participation in informal exchange was part of what it meant to be a member of the community; only in this particular case, the community was German. In contrast to Anglo areas, interhousehold reciprocities were never overwhelmed by the staples economy and continued to order and moralize what would otherwise have been an increasingly amoral and market-defined world. Karl Urbantke, who came to Austin County in the 1850s, emphasized the vitality of interhousehold labor arrangements in Cat Spring, where it was "the custom in our community to [help build houses and barns] on Sunday. The neighbors gladly came and helped for there were always good things to eat and drink."[38]

In other respects German sociability differed from that in Anglo areas. Whereas some Anglo communities, such as Chappell Hill, were knit together in evangelical Christian "brotherhood," formal religion had a much more muted presence in the German communities. The reasons are far from clear. Some of the immigrants were secular *Lateiner*, or university-educated liberals, republicans, and romantics who eschewed formal religion. The older settlements in Austin County appear to have had more than their share of professional and well-educated residents. Among these was A. F. Trenckmann of Millheim, who as a self-described "Christian Catholic" opposed both "ultramontanismus" and "pietitiszmus" in an effort to "harmonize belief and science." The presence of Lateiner like him may partially explain the absence of a Lutheran church in such a heavily German area.[39] Even so, such immigrants could have comprised only a small part of the population. Surviving letters demonstrate that religious feeling was strong among Germans, although it seems to have been expressed privately or in informal settings. In 1859, Wilhelm Schlottmann wrote from Brenham that he and his neighbors had built a church, which would also serve as a school. "You must not, however, imagine it to be a great palace," he told friends back in Wehdem, "and we have no organ or steeple. This would have been too much for so few people."[40]

German communities were also knit together in formal organizations and societies, which served not merely to aid fellow immigrants and foster ethnic solidarity, but also to mitigate the comparative sparseness of New World settlement. In this respect the Germans resembled Anglo Americans, although their organizations tended not to overlap. The oldest and most important organization was the Agricultural Society of Cat Spring, which was founded in 1856. The official purpose was "solely to develop an interest in agriculture and to solve related problems," and to that end it offered community subscriptions to newspapers and agricultural periodicals in both German and English, which some members translated for non-English-speakers. The organization also provided modest economic benefits to members. During the drought of 1856–57, it persuaded a local merchant and gin owner who belonged to the organization to gin members' cotton at a reduced toll.[41]

Organizations like the Agricultural Society of Cat Spring also bound area Germans together in less formal, but equally important ways, all of which contributed to the construction of a separate German identity. The society periodically staged community celebrations and "frolics," which were open to members, their families, and nonmember friends. Karl Urbantke described an event in Millheim in the 1850s: "A quartette gave an admirable rendition of German songs. Festivities were often lauded in poetry, . . . champagne usually found its way into these gatherings, and wit and wine flowed freely."[42] It also staged July 4 gatherings and purchased a U.S. flag to be hoisted at the annual celebration, a gesture that sought to harmonize German with "American" identity.[43] For women, whose duties generally confined them to the farmstead, society gatherings broke down rural isolation and connected them to the larger community, since, as Charles Nagel explained, women "did not have occasions for meetings like the men."[44] Such events ensured that a sense of German identity and community would extend beyond the walls of the society's meeting hall, instilling it in men, women, and children of all ages.

As a result of such highly elaborated community life in their respective neighborhoods, interaction between Germans and Anglos was generally restricted to formal settings, such as politics or the marketplace. Rarely did Germans and Anglos venture into each other's churches or meeting halls. The Germans had, for the most part, succeeded in drawing a boundary between themselves and the Anglo plantation world. It was a boundary that became increasingly troubling to Anglos as the 1850s progressed. Over time, German ethnicity was mapped according to the community's public, political stances, rather than in reference to their private, familial practices.

Among the more thorny issues between Germans and Anglos was alcohol, particularly in connection with the Sunday events so common in the immigrant settlements. Evangelical Protestants were generally horrified at the thought of Germans drinking and dancing on the Sabbath and registered their displeasure numerous ways. Lem Swearingen of Millheim, whose children spoke German and who enjoyed otherwise good relations with his European neighbors, lobbied the Agricultural Society of Cat Spring to reschedule a July 4 feast from Sunday to Monday, presumably to avoid profaning the Sabbath.[45] Whether the society did so is not recorded, but generally speaking, the immigrants paid little heed to Sabbatarian complaints. German drinking also surfaced as a political issue in 1854, when an initiative appeared on the Texas ballot that would have given localities the option of banning the sale of liquor in quantities of less than one quart. The measure failed miserably throughout the state, including the counties along the Brazos, where it garnered only 12 percent of the vote in Brazoria County and 28 percent in Fort Bend County. In Austin County, however, it received 49 percent, which suggests that immigrant practices had made many Anglo voters (who still comprised a majority) uneasy enough to support restrictions.[46]

Ultimately, it was the image of the immigrants as unsympathetic to slavery that emerged as the most critical issue between Germans and their Anglo neighbors. This image was by no means confined to the Anglo community. In 1938, ex-slave Sarah Ford, who had lived on the Patton plantation in Brazoria County, told a WPA interviewer about her father, Mike, who as a runaway had been shielded by a neighboring German family. Ford gave the German family's name as "Ebbling," almost certainly Charles Eberling of Columbia. A tanner by trade, Mike bartered his skills for shelter. To Sarah Ford, the Eberlings were "white folks what don't believe in havin' slaves . . . some German folks."[47]

Although there was a common perception that the immigrants opposed slavery, it was by no means universally true. If pursuit of the Main Chance could lead to widespread slaveholding among Anglos, it could do so among the profit-minded immigrants as well, and some did indeed buy slaves. Charles Amsler, a German-speaking Swiss who moved to Austin County with his brother and their families in 1834, purchased his first slave in 1845, and by 1860 he owned seven. The slaves tended his cattle, ran his mill and gin, and tilled his cotton fields, which yielded thirteen bales of cotton on the eve of the Civil War. According to his grandson, Amsler bought laborers for no reason other than "to have help to make more [money]" and had little problem with slaveholding because he and his family "claimed to be no better than their neighbors hence could see no wrong in doing as the majority did."[48] Franz Hinze bought a slave less than two years after coming to Independence, Washington County, from Varl. In his first year, he parlayed a stake of three hundred dollars into a three-hundred-acre farm. The next season he hired eight laborers (almost certainly German) and managed to produce thirty-five bales of cotton. Hinze sold the cotton for $1,500 and made another $1,000 with his mill and gin. After expenses, he had $1,750 left over, $1,300 of which he spent on a slave. Writing to friends back home, he explained, "I have bought myself a black man who will be mine as long as he lives, and when I die my children will inherit him." Moreover, Hinze appears to have had few qualms about working his slaves hard, as he went on to tell his friend that a single slave was "worth as much as two servants [*Knechte*]," and that "here the black people are all slaves who can be bought and sold. They have no rights in relation to whites, [and] no whites can be sold."[49]

The experiences of Amsler and Hinze aside, the vast majority of Germans did not own slaves. Whether it was due to economic or cultural factors has long been a matter of debate. Geographer Terry G. Jordan has shown that German farmers in the "eastern German belt" on the Brazos and Colorado rivers displayed an orientation toward cotton and corn culture that the immigrants in the more homogeneous "western German belt" in the hills north of San Antonio did not. This pattern suggests that the Germans might have become slave owners given time but did not because they lacked the capital, having only recently paid for passage to Texas and their farms. Whatever cultural biases Germans had about owning slaves were not a significant hindrance, Jordan argues, but economics and political circumstances were: the Civil War ended slavery before most Germans could afford slaves.[50] In a similar vein, it is also worth noting that a steady stream of new immigrants in the 1850s reduced the need for slave labor in German communities. Germans not only rented land from older residents; they also constituted a fairly large pool of wage labor, providing German farmers with a cheap resource that was, with occasional exceptions, unavailable to Anglos. Eventually, most immigrants succeeded in buying land for themselves within a few years, but the flow of newcomers remained steady enough to supply older residents until the Civil War.

Although there can be little doubt that a lack of capital did prevent some Germans from buying slaves, whether that alone accounts for the immense gap in slaveholding, or whether ethnocultural biases were responsible is difficult to say. Historians who argue that German misgivings about slavery were significant in keeping them from buying bondsmen generally point to the same statistical evidence that their opponents do. In Austin County in 1860, eighteen Germans owned a total of seventy-eight slaves; these numbers represent about 5.6 percent of all slave owners and 2 percent of the slaves in the county.[51] In addition, argues historian Walter D. Kamphoefner, the Germans' preference for purchasing land over slaves demonstrates that they possessed the necessary capital but consciously chose to invest it in something other than slaves. The evidence for western Austin County bears this view out. The ratio of real to personal property in 1860 was 1.4:1 among Anglos and 1.8:1 among Germans, compared with .98:1 for the entire Brazos region.[52]

Unfortunately, the qualitative sources that can best answer the question rarely mention slavery at all, but that silence in itself may be the most significant statement on the matter. The Brazos Germans never denounced bondage explicitly, as the German editor of a San Antonio newspaper did in 1854 and was driven out of the area for it. Of course, it is important to remember that the Brazos was a plantation region where slavery was the central economic institution, whereas San Antonio was an urban area located well outside the plantation belt, where slavery was not so important. Even the greenest newcomer to the Brazos could have sensed that open criticism was probably a dangerous business. In sum, given the range of factors that may have influenced slaveholding, ownership and property statistics cannot actually prove an ethnocultural hypothesis; they merely do not disprove it. The question of whether the glass is half full or half empty remains.

Perhaps a more useful way to think about the question of immigrants and slavery is to set aside the matter of whether the bulk of Brazos Germans consciously opposed the institution. Clearly some did and some did not. We might instead ask how the Germans made sense of the historically and culturally specific variety of slavery they encountered. After all, foreign migrants to the Brazos encountered not a Platonic ideal of slavery, but an institution that was, as all slave societies were, the product of specific circumstances. This particular variety of slavery was shaped primarily in reference to a culturally specific language of family. Anglo family members, as we have seen, spoke to one another in the language of affection. "Correction" and "chastisement" were justified by the supposed love and concern a master had for the well-being of his dependents. Evangelical churches blended a message of affection and submission on the basis of race and gender to lend it moral support and legitimacy. Finally, as we have seen, Anglos regarded family-based slavery as the proper way to civilize a savage wilderness.

Most new immigrants, however, were not conversant in the idiom of late-antebellum borderland paternalism, and a host of imperfect translations ensued. Put differently, paternalism had acquired various articulations with ideas and cultural practices, such as familial affection, evangelical religion, and the imperative to order wilderness chaos. Affection was without question important in German families, but it was not charged with the same ability to compel as it was among Anglos. And whereas evangelical Protestantism offered support for bondage, immigrants tended to regard Anglo religion with circumspection, even suspicion. For Gustav Dresel, the emotionalism of Texas Protestants rang hollow. Of one preacher he wrote, "I had never found such a mocker of the better feelings of humanity and it infuriated me."[53] Instead of seeing slavery as a way to civilize a savage wilderness, Karl Urbantke saw Anglos as rowdy and slaves as wild men who consumed raw ears of corn and "crushed the kernels with their teeth and ate them with a little molasses."[54] With that in mind, it is worth asking whether much of Anglo slave society might have appeared, in the absence of unusual perceptiveness on the part of the immigrants, rather alien and inexplicable. If so it would have been difficult for new arrivals to adopt Anglo views of slavery without questioning many of the ancillary concepts that ordered their own view of the world. The question is not whether the Germans approved of slavery; rather, at issue is whether they were able to accommodate it easily given their understanding of family and power. Just as the Africans of Brazoria had to learn to speak the paternalist pidgin, so did the Germans. Some embraced the Anglo emotional vocabulary, but most would have done so only gradually and fitfully, and many perhaps never.

Complicating matters is that Germans did not arrive tabula rasa in matters of subjection. Although it was the subject of immense political dispute at the time, most Germans were quite familiar with *Herrschaft*, or domination. One historian has proposed that familiarity with feudal Herrschaft may have facilitated the adoption of "pro-slavery" attitudes on the part of some elite immigrants.[55] Another has argued that German women viewed and treated their female slaves as European-style servants.[56] There may indeed have been some for whom Herrschaft translated into mastery, although the vast majority of immigrants certainly did not come from the ranks of the aristocracy. Those Germans who compared serfdom with slavery often found more differences than similarities. As we have already seen, when slaveholder Franz Hinze compared his own bondsman in Texas to a German Knecht, he stressed that he was able to extract twice as much labor from the slave. Hinze also, probably in response to queries from back home, found it necessary to clarify the legal standing of Negroes in Texas. Unlike Knechte, he informed his correspondent, blacks in Texas were slaves for life and "can be bought and sold. They have no rights in relation to whites, [and] no whites can be sold."[57]

Travelers' guides and popular literature also helped ensure that immigrants arrived in Texas with some notion of slavery and race. Viktor Bracht's popular guidebook informed its readers that slaves were "well treated, and, aside from the thought of perpetual slavery, their lot is undoubtedly far better than that of many servants and most factory workers of Europe."[58] In fiction, the works of Friedrich Gerstäcker were especially popular, and his name was even mentioned by Brazos Germans as a favorite.[59] Gerstäcker, who traveled through the United States in the 1840s and 1850s, produced several novels about American life, filled with romance, adventure, and above all, descriptions of Americans that appealed to the same curiosities as did works by Alexis de Tocqueville, Frances Trollope, and other European observers. And Gerstäcker's Americans were much like those depicted by his contemporaries, especially in their single-minded pursuit of profit. Gerstäcker, however, had a special appreciation for what he called *Die Backwoodsmen Nordamerikas*, who represented "the opposite of bourgeois life" and clearly owed a great debt to James Fenimore Cooper's *Leatherstocking Tales*.[60] But it is his depictions of African Americans that are most interesting. Gerstäcker's black characters were often dangerous, and never more so than when they seemed to be out of their proper place. His *Wanderings and Fortunes of Some German Immigrants* (1848), in which a party of migrants settles in Tennessee, supported a variety of stereotypes: black stevedores attempt to steal the immigrants' luggage in New York; black muggers assault the main protagonist on his first night in America; a black barber in Cincinnati spooks his German customer when he picks up his razor; and the principal villain and his free mulatto henchman kidnap the beautiful daughter of a Lutheran minister. The only sympathetic African American character is a faithful slave who belongs to a kind-hearted German woodcutter; all the other black characters are free and disorderly.[61]

Another prolific German novelist was Karl Postl, who wrote similar stories of adventure in American settings. One of his stories, *Das Cajutenbuch*, or *The Log Cabin Book* (1847) was actually set in Texas at the time of the War for Independence, and it borrowed freely from the anonymously written 1831 travelogue *A Trip to Texas*. Unlike Gerstäcker, whose authorial voice was that of a European familiar with Americans, Postl effectively became American in his novels, which were published under the pseudonym Charles Sealsfield. Postl's views on slavery differed little from those of most of the Anglo South. In *Der Pflanzer am Red River*, whose story concerns the creation of a plantation on the Red River Frontier of Louisiana, Postl enthusiastically embraced the entire proslavery argument and transmitted it to his German readership.[62]

Finally, German intellectuals in the early nineteenth century had begun to confront slavery and race, and their ruminations spilled into political discourse as the century progressed. At least since the formal abolition of serfdom in 1807, Germans had been grappling with questions of equality and freedom, a process that culminated in the events of 1848. For German intellectuals and politicians, slavery became a touchstone in the debate on the merits and defects of particularism, estatism, and liberal economics. As elsewhere, opinion covered a broad spectrum. Conservatives found little difficulty with slavery as they defended the lords' privileges in regard to their servants. Liberals were divided. Some opposed slavery as anachronistic, inefficient, and a violation of the rights of man. Others, while perhaps sympathetic with these points, nevertheless were reluctant to support abolition because to do so would violate the owners' property rights.[63] With works like Gerstäcker's circulating among Germans in Europe and the United States, and intellectual discussion addressing slavery, immigrants had had a great deal of prior exposure to the debates over slavery and race.

Not surprisingly, then, slavery and race ordered the Germans' own Brazos settlements in unexpected ways. Charles Nagel recalled playing a game, as a child, with other German children called "Black Man's Out," which he insisted had nothing to do with race because "black men were almost unknown to us."[64] Yet it was on Nagel's father's farm in Millheim that fellow immigrant Karl Urbantke was taught the connection between race and equality. Arriving in the mid-1850s, Urbantke, like so many others, worked as a laborer for a family that had arrived earlier. One of the older laborers at Nagel's farm quickly advised him not to allow himself to be treated "like a nigger." When Urbantke asked for clarification, the laborer told him how he had refused to care for Nagel's horse after a long day's work. Although Urbantke remarked cryptically that such insubordination was "not always practicable," he came away from the conversation with the impression that "on the social ladder both employer and employee stood on the same rung. They ate at the same table and shared their joys and sorrows."[65] At least for the laborers on Nagel's farm, a moral economy of race prevailed that differed little from that of the Anglo South.

In the final analysis, we must return to Sarah Ford's characterization of Germans as "folks what don't believe in havin' slaves," which speaks less to the attitudes of the immigrants toward bondage than it does to the more important question of the meaning of Germanness in a slave society. What Ford perceived so clearly was that slavery as practiced in the late-antebellum South made little cultural sense from the German perspective. Affection did not imply submission. God did not ordain the eternal subordination of the descendants of Ham. Civilization was not inseparable from slavery. Of course, individuals were free to ignore or adapt any of those propositions, and the situation in the immigrant communities was far from static. The fact that it was generally the older residents who owned slaves certainly suggested that acculturation and greater economic resources could combine to change attitudes. But what immigrants could not ignore was that the significance of Germanness on the Brazos was not wholly within their control. Anglos and African Americans both had their say, with sectional politics providing the setting for the conversation.

"The Black Dutch"

The late 1840s and 1850s represented a departure in Brazos politics. Border politics reached a crisis point in the 1840s. In 1841, President Mirabeau B. Lamar antagonized Mexico by dispatching an expedition to Santa Fe in hopes that residents could be persuaded to sever ties with Mexico and by supporting a federalist rebellion in the Yucután. Mexico, which did not recognize Texas independence, responded in 1842 by briefly capturing San Antonio. The Republic retaliated by sponsoring a raid across the Rio Grande, which was led by Arthur Somervell, who had been James F. Perry's business partner in San Felipe before Perry moved to Gulf Prairie. A number of Brazos residents flocked southward to defend the border claims of the new republic, including at least twenty from Brazoria County alone. After reaching the Rio Grande and taking Laredo and Guerrero, Somervell recognized that he was in a vulnerable position and ordered his troops to withdraw into central Texas. Most of his force, however, refused to do so and continued on to plunder Mexican towns along the river, electing a new leader, William S. Fisher, to replace Somervell. The Mier Expedition, named after a town on the southern bank of the Rio Grande where the raiders had tried to obtain supplies, ended in disaster. A Mexican force captured the raiders after a short battle. The prisoners were then marched to Mexico City, where it was eventually decreed that one of every ten should be executed. The method of selection was to mix a small number of black beans in with white beans. Those who drew black beans were put to death.[66]

The Mier Expedition underscored Texas's vulnerability. The new president, Sam Houston, favored annexation to the United States, but opposition by northern Whigs complicated matters. The alternative to annexation was to continue existence as an independent republic. This choice would likely have required courting Great Britain as a financial and political patron. British patronage would have come with strings attached, possibly even a compensated emancipation. Or at least that is what Houston let U.S. officials believe. In 1845, however, the newly elected and avowedly expansionist Polk administration annexed Texas, and with the military victory in 1848 in the Mexican-American War, it managed to force Mexico to accept that Texas was part of the United States. The new international boundary was set at the Rio Grande.

With the disappearance of independence as a political issue, a semblance of a two-party system emerged. Its operation, though, was curtailed by the crumbling of the national parties under the strain of sectionalism. In fact, Texas's admission to the Union was in itself a significant source of sectional tension. Throughout the life span of the second party system in Texas, the Brazos was overwhelmingly Democratic. This was the case despite the growing number of wealthy planters, including tariff-dependent sugar planters, who in other states made up an important constituency of the Whig Party. In every gubernatorial and presidential election between 1848 and 1857, regular Democratic candidates defeated their opposition handily in the lower Brazos counties. The exception was Opposition candidate Sam Houston, who won the governorship in 1857. Houston received only 42 percent of the vote along the Brazos, largely as punishment for the vote he had cast in the U.S. Senate against the Kansas-Nebraska Act. Other than that, Brazos politics in the 1850s seems at first glance to have been driven by consensus on most issues.

The Democratic winning streak, however, obscures the fact that the politics of neighborhood and personality that had existed only a few years before collapsed in the 1850s. As the decade progressed, support for Democratic gubernatorial and presidential candidates along the Brazos declined from a peak of 77 percent in 1851 to a slim majority of 52 percent in 1855 (see table 10). In the hotly contested gubernatorial race of 1859, the winning candidate, Sam Houston, received only 50.1 percent of the vote. Most striking is the strength of the American, or Know-Nothing Party in comparison with the Whigs. Although the Know-Nothings contained many ex-Whigs, it enjoyed stronger support on the Brazos than the party of Henry Clay had, largely because of its nativist appeal. County returns support the point: in the 1855 gubernatorial election, Know-Nothing David C. Dickson garnered 38 percent of the vote in Austin County and 48 percent of the vote in Washington County. These two counties had the largest immigrant populations along the Brazos, though in neither case were immigrants a majority. The relatively strong support enjoyed by the American Party was in all likelihood due to Anglos' suspicion of their German neighbors. In contrast, Brazoria and Fort Bend counties, neither of which had a significant number of German voters, each favored Democrat Elisha M. Pease, a Brazoria County resident, by 88 percent. The Whigs had never won more than 25 percent.[67]

Another trend illustrates the growing political split of the 1850s. In every election through 1852, there was little difference between Upper and Lower Country voting patterns. Beginning in 1853, Washington, Grimes, and Austin counties supported different candidates and different policies. Much of this change had to do with the large immigrant population centered in Austin County, which made the American Party particularly popular. But there were issues among Anglos as well, such as the Texas debt. In 1855, after a lengthy dispute over whether the federal money received in the Compromise of 1850 should fund the debt at par or at a "scaled" rate, the legislature approved the former course. Those who were most dependent on credit, such as merchants and many planters, favored the plan. Galveston and Houston newspapers even linked funding to "honor."[68] In the Upper Country, where the growth of staples agriculture had been more rapid, the measure was far less popular. As Anderson resident and gubernatorial candidate David C. Dickson expressed it, funding was desirable in the abstract but "ignores all equity in its practical application."[69] In Fort Bend and Brazoria the plan received 59 percent support. The Upper Country voted it down, however, with Grimes, Washington, and Austin counties giving it 26 percent of the vote. The measure lost statewide as well, but since the referendum was nonbinding, it remained the law.

As the 1850s progressed, sectional tension loomed larger. At times, issues that had little to do with sectionalism became entangled with it, as the railroad did in the mid-1850s. Although some small railroads already linked the Brazos with Houston and Galveston, the state took up the question of how to fund a more comprehensive system. Promoters from Houston and Galveston soon inaugurated campaigns to make their respective cities the system's primary terminal. The Galveston-centered, or State Plan, envisioned a state-built and state-owned system. The Houston-centered Corporate Plan proposed to aid private corporations with the loan of public land from the state's school fund. Debate was fierce, but in 1857 the Corporate Plan won. Charges that the principal author of the State Plan was an abolitionist clinched the legislature's decision. Sectionalism, it seemed, could forge consensus in a society rent asunder by credit and markets.[70]

Feelings of neighborliness could also be revived by facing a common threat. Yankee abolitionists fit the bill at the national level, while German immigrants could be cast as their agents at the local level. Leading the way on the issue was the American, or Know-Nothing Party. The Know-Nothings appealed to former Whigs who could not tolerate merging with the Democrats. The party portrayed itself as the only organization capable of ensuring the survival of slavery by keeping antislavery agitation out of politics and focusing instead on the problem of foreign immigration. Anti-immigrant politics were most apparent in the columns of the *Washington American*, the local organ of the Know-Nothing Party, which worked tirelessly to associate immigrants with antislavery. For several weeks it ran the texts of antislavery declarations passed by German associations in Virginia and Massachusetts. One article on a speech by Sam Houston, who eventually allied with the Know-Nothings after being forced to the sidelines of the Democratic Party by the Southern Rights wing, exemplifies the effort to connect immigrants with antislavery.[71] According to Houston, the Democrats were responsible for "the promiscuous introduction of the elements of abolitionism, by encouraging the promiscuous introduction of foreigners, in direct conflict with the best interest of the South, of the perpetuity of the Union, and the safety of the institutions of freemen."[72]

Fear of German abolitionism increased when Adolph Douai, editor of a German newspaper in San Antonio, published the resolutions of an antislavery convention held in that city in 1854. The statement simply confirmed what many Anglos along the Brazos felt they already knew—European immigrants were a natural threat to their domestic institutions. "It is absolutely preposterous for men to talk about loving an institution, which they have been taught to abhor from their very infancy, until practical demonstration convinces them of their error," maintained the editors of the *American*, "Hence it is that foreigners who have settled in large bodies in the country, keep up this feeling of antipathy to our institutions." The editors, however, took pains to emphasize that not all Germans were the same. Those in the Hill Country north of San Antonio were almost certainly irreconcilable abolitionists because "they are settled to themselves, keep to themselves, speak their own language, refuse our teachers and preachers, know nothing of the opposition [to] slavery, and hence oppose it." But the editors held out hope that Brazos Germans might not be hostage to the same prejudices. "With those who have scattered off through the country, and settled in small bodies, or to themselves, the case is entirely different. They see and appreciate the system, and as soon as they can, become slave holders themselves."[73] The latter statement, of course, only dimly reflected reality on the Brazos. While the German settlements in the Hill Country were certainly larger and more homogeneous, those in Austin and Washington Counties were, as we have seen, neither scattered nor small in number, and slaveholding, for whatever reason, was nowhere nearly as prevalent as in the Anglo-dominated areas. As immigration increased in the 1850s, it grew more difficult for Anglos to convince themselves that the Germans were a benign presence.

This fusion of nativism and sectionalism posed a subtle threat to the plantation hierarchy. In particular, given the importance of race in justifying slavery, associating the Germans with antislavery endangered the construction of the black-white binary. The epithet Black Republican, ubiquitous in the southern press at the time, suggested that those who opposed slavery were not quite white themselves. In fact, use of the term among Irish immigrants in the North was an important rhetorical strategy designed to bolster their status as whites.[74] On the Brazos, the appearance of similar language marked the subtle racialization of Germans as black, at least among the Know-Nothing faithful. One article in the *Washington American* referred to Austin County as "the county of Dutchdom" and "Frankfort-on-the-main," inhabited by "the black dutch."[75] Another article on a political meeting in Bellville insinuated that Anglos who associated with the "Kroutocracy" of Austin County were of a color that "shall be here undescribed."[76] The strategy, while politically expedient in the short term, ultimately up-ended the notion that all whites were equal, which in turn undermined the basis of nonslaveholding whites' support for slavery. To be certain, few on the Brazos were seriously inclined to reclassify Germans in the grand taxonomy of God and nature. Still, such talk contributed to the construction of Germanness as an oppositional identity and the immigrants' image among blacks as "white folks what don't believe in havin' slaves."

As the Kansas crisis and the issue of slavery in the territories eclipsed all other concerns in 1856–58, the Know-Nothings faded. John Brown's 1859 raid on Harpers Ferry fed fears that abolitionists were determined to foment servile insurrection in the South. Then, in the summer of 1860 a rash of fires aroused suspicion that a similar plan was afoot in Texas. The first fires broke out on July 8 in the north Texas towns of Dallas, Denton, and Pilot Point, where temperatures had reached as high as 110 degrees. Dallas in particular suffered significant damage, losing half of its downtown district. In the next few weeks, more fires broke out across the state. Although it was later determined that the original fires were probably caused by a new type of match that was prone to spontaneous combustion in exceedingly hot and dry weather, whites soon concluded that the fires had been deliberately set. Alleging an alliance between slaves and abolitionist preachers to sow chaos as a prelude to a massacre of white men and the wholesale rape of white women, whites began rounding up suspicious slaves. An intercepted letter from a Methodist minister supposedly confirmed the diabolical scheme. Vigilantes kidnapped the minister in Missouri, transported him to Texas, and executed him. In addition, at least thirty, and possibly as many as one hundred, slaves were hanged. Although most of the action took place hundreds of miles north of the Brazos, rumors of incendiaries in Matagorda County, Brazoria's western neighbor, set local nerves on edge. The hysteria was fed by fears that the Republican Party would win control of the federal government in the fall elections. Democratic politicians exploited the hysteria in turn, pointing to the "plot" as evidence that southerners needed to back the solidly proslavery John C. Breckinridge in the coming presidential election. Their opponents, the Fusionists, charged that the Democrats were exploiting the fear of insurrection for political gain, but the allegations had little effect.[77]

By election time in 1860, the future seemed almost foreordained. At the Democratic convention in Charleston, delegates from the lower South, including convention chair Guy M. Bryan of Brazoria, along with five other Texas delegates, walked out over the party's failure to endorse a national slave code for the western territories. Bryan's constituents generally approved of his actions and supported the candidacy of John C. Breckinridge, who was nominated at a subsequent convention on a platform that called for the federal government to ensure that "all citizens of the United States have an equal right to settle with their property in the Territor[ies]."[78] A victory by the Republican Party, most understood, would signify the permanent helplessness of the South before what John C. Calhoun had years before called "King Numbers," and would result in disunion. "The people are running wild upon the abolition question and if Abram Lincoln is elected President I think we will have some distressing times this winter," wrote James McKenzie, who as a nineteen year-old law student had only just arrived at Brazoria from Kentucky. "I say Hurrah for Breckinridge & Lane," he wrote, "and I believe the state of Texas will give its voice for them [in the] election."[79]

McKenzie was right. Brazos voters favored Breckinridge by a ratio of four to one. Only 20 percent of the ballots went to the "Fusion" or "Opposition" ticket, a coalition of electors who pledged to support whichever candidate appeared to have the best chance of defeating Lincoln. Not surprisingly, the German areas delivered much of the anti-Breckinridge vote. In Cat Spring, Fusion/Opposition electors received thirty-two votes, against eight for Breckinridge. In Shelby, another German neighborhood in western Austin County (also known as Roeder's Mill), the Opposition won 64 to 13. In Washington County, where the German population was smaller, Fusion candidates failed to carry a single district but fared respectably in some of the western precincts. In Union Hill, for example, they received 34 percent of the vote.

But Breckinridge carried a few heavily German districts by a considerable margin, the most remarkable being the 76-to-5 vote in Industry. The contrast between voting in Industry, on the one hand, and Cat Spring or Shelby, on the other, is difficult to explain. All were overwhelmingly German. Cat Spring and Industry were both older settlements, each with a few longtime residents who had acquired slaves. Although some contemporaries suggested that Cat Spring and Millheim had proportionally more liberal Lateiner, there were hardly enough to make a significant difference. One historian has argued that politics in Industry were dominated by a particularly powerful clique of longtime German residents who worked hard to deliver the precinct to the regular Democrats. It is very possible that older, wealthier residents used ties of patronage to get tenants and hired laborers to vote for Breckinridge, although there is no direct evidence for it. It may also be the case, as it was in New Braunfels in the Hill Country, that community leaders counseled voters not to provoke a reaction when Breckinridge seemed sure to win. In any event, the totals at Industry seem anomalous in comparison with those of Cat Spring and Shelby, but the precise reasons for the election outcome are likely to remain obscure.[80]

The Fusionists also did reasonably well in the region's towns, which contained a disproportionate number of immigrants, northerners, and merchants, who generally favored more moderate measures. In the town of Chappell Hill, which had a small population of immigrants, 34 percent of voters supported the Fusionists, while in Bellville approximately one in five voters supported the ticket. They also did slightly better in the Upper Country than in Fort Bend or Brazoria, where Breckinridge garnered majorities of 90 and 84 percent, respectively.

Of course, although Breckinridge carried Texas and the rest of the lower South, Lincoln won nationally, precipitating the secession crisis of 1860–61. Secessionists in several towns along the Brazos organized public meetings to select delegates for a convention whose mission would be to vote on disunion. One such mass meeting occurred at the courthouse in Brazoria on November 17. Those in attendance selected a Committee of Safety and four delegates to go to Austin "to take into consideration, the extraordinary state of affairs, likely to arise out of the recent election of a Sectional Candidate for President of the United States by a Sectional Party averse to the Constitutional rights of this and the other slave holding States."[81]

The politics of secession in Texas soon grew rather complex. Sam Houston, opposition leader, Unionist, and now governor of Texas, considered secession unconstitutional and refused to convene the legislature, which was the only body with the authority to call a convention. Secessionists then called one on their own on the pretext that the circumstances permitted a sovereign people to take up matters of such importance whether the legislature was in session or not.[82] To nobody's surprise, on February 1, 1861, the convention approved an ordinance of secession, explaining that the federal government "is sought to be made a weapon with which to strike down the interest and prosperity of the people of Texas and her sister slaveholding states."[83] The articles also mandated a referendum on secession, which was held on February 23, 1861.

The vote in favor of the ordinance was overwhelming. On the Brazos, only Austin County demonstrated any opposition at all, with 26 percent of all voters rejecting secession. According to William Trenckmann, some of the older, slave-owning families in Millheim, including the Klebergs, the Langhammers, and the Engelkings, supported secession. They "had been in the country so long that to them as to almost all Anglo-Americans slavery in the South seemed unavoidable and therefore justified." On the other side, there were a few Lateiner like Herman Nagel and Adelbert Regenbrecht who "enthusiastically supported the freeing of the slaves" and were outspoken at least locally in their opposition to secession. Debate between the two groups was heated. Most Germans, however, probably regarded secession as an unwelcome intrusion, but aware of their delicate position on the Brazos and "expecting nothing but harm and certain defeat" from conflict with their Anglo neighbors, stayed generally silent.[84]

Along the Brazos, the German areas of Austin County exhibited the only significant opposition to secession. Only nine votes in Grimes County were cast against disunion, and Fort Bend registered not a single vote against. In Brazoria, where secessionists hovered at the polls and passed out ballots marked "For Secession" (and presumably watched the voters drop them in the ballot boxes), only two votes were cast against secession.[85] These totals strongly suggest voting irregularities, possibly fraud, but more likely intimidation and a sense of futility on the part of Unionists.[86] Undoubtedly many Unionists, glimpsing the writing on the wall, simply stayed home, although turnout was actually higher for the secession referendum than for the presidential election. Still, the absurdly low Unionist turnout suggests that there were many who felt as Oscar Addison did. "Dissatisfied as I have been with the cause of the Southern States," he wrote, "I could not vote for Secession [and] Accordingly I did not vote."[87] In the end, however, Lincoln and the Republican Party appeared to have accomplished what other measures could not—the veneer of consensus among neighbors on the Brazos.

For longtime residents of the Brazos, secession was nothing new. Only twenty-five years before, Texas had seceded from the Republic of Mexico. In both 1836 and 1861, Anglo residents concluded that the national state posed a threat to slavery. The most obvious difference was that in 1836 (and again in 1845), Anglos judged slavery to be safe within the federal system of the United States. By 1861 that was no longer true. A president had been elected without a single southern electoral vote. For most whites in the lower South, Lincoln's election heralded a new era of political domination by the North, one whose outcome seemed almost too nightmarish to ponder. At the same time, much had changed at the local level. In 1845, the neighborly reciprocities that sustained consensus on matters pertaining to household and family were in good order. By 1861, however, the growth of the staples economy had left many Anglos feeling more alienated from their neighbors. That, coupled with the growth of the German settlements, inspired much nervous talk of the "Black Dutch."

The crisis in neighborly relations was brought about by sectionalism. Although it was not apparent at the time, the most significant result of neighborhood politics in the 1850s was the construction of Germanness as an oppositional identity. Anglos contributed unwittingly to this point of view by making the Germans a foil for their own efforts to recreate an imagined political community. The Germans themselves contributed by constructing an ethnic identity that was distinct and separate from the Anglos. Lastly, African Americans contributed by investing Germanness with an unequivocal antislavery cast, deserved or not. In many ways, they had done the same in regard to Mexico more than two decades earlier. Even slaves in the late antebellum period viewed the Mexican border as a place beyond which slavery was invalid and where racial categories were more fluid. Now, as Sarah Ford's narrative shows, slaves discerned a similar boundary between Germans and Anglos, one that soon threatened the hierarchies of Brazos plantation society.

Chapter 5

WAR

Thomas Blackshear, owner of one hundred slaves and a plantation near Navasota in Grimes County, was a man of few words. On February 23, 1861, he made a typically brief entry in his pocket diary: "Trying to rain again. Plum and peach trees in bloom. At 11 oclock AM a Norther sprang up and blew coldly until night. Brought all hands from the bottom. Finished building negro houses at the bottom." At the back, however, he recorded an uncharacteristically telling thought, indicating that today was not like other days. "The people vote today to Ratify the action of the convention in passing an 'ordinance of Secession' from the United States," he wrote, "God grant it may be best for the people, but I am fearful of trouble."[1]

Elsewhere in the South, the news of secession was greeted with fanfare and cheer. Along the Brazos, despite the endorsement of a majority of the voters, the mood seems to have been more somber and apprehensive. There were several good reasons for this. The Brazos had enjoyed ten years of spectacular economic growth. The amount of cotton produced in the five counties of the lower Brazos had increased more than 400 percent. The population had increased 130 percent, driven by a 140 percent increase in the number of slaves, who now made up 55 percent of the total population. Now, with the prospect of being cut off from markets and eastern credit, Texas planters like Blackshear had good reason to be worried.[2]

Secession euphoria was dampened by another fear, that of slave rebellion. In the town of Columbia, the rumor circulated that an uprising was planned for March 4, 1861, Lincoln's inauguration date. The "proof" was the alleged deathbed confession of a slave girl in neighboring Matagorda County, who was said to have outlined an abolitionist-inspired plot in great detail, although nobody seemed to know the particulars. Whites in Columbia braced themselves for the worst, patrolling the town with heightened vigilance and ordering all slaves to stay at home. Oscar Addison, who had hoped to preach to a slave congregation on the evening of the fourth, was incredulous. The sketchiness of the plot made him suspect the it was the work of "wise acres," "designing bad men," and a "partizan press," possibly for political reasons. In fact, Addison thought, loose talk of rebellion was more harmful than anything the slaves might have planned. "In my estimation," he confided in his journal, "the whites by their talk, (which many of the negroes hear), and manner of doing in the affair, has a legitimate tendency to create the very thing they fear, the existence of which, is confined to the fears."[3]

Addison went out on the evening of March 4 to keep his preaching appointment. Only four people came to hear his sermon. As he returned through the muddy streets of Columbia to his hotel, a bell began to ring, the signal for all whites to take shelter in the town jail. The bell, it turned out, was a false alarm, and the evening passed uneventfully. As far as Addison was concerned, the people of Columbia had acted "old womanish," not only in their fear of revolt but in their choice to secede. The entire situation seemed the work of "demagogues" who, with stories of "abolition pests, poison and arson" had "inflame[d] the public mind." Lincoln, after all, had been elected constitutionally. Why "it should be urged as sufficient grounds for the disruption of one of the best governments under heaven" was beyond him.[4]

Although they could hardly have known it, Addison and Blackshear were witnessing the beginning of the end of slavery, and with it the passage of the Brazos from borderland to bordered land. Ironically, with the start of the war, the pull of Mexico on the Brazos was stronger than at any time since the signing of the Treaty of Guadalupe Hidalgo. With its own ports blockaded, planters drove their crops over the border to Matamoros, whence they were then shipped to European markets. Needless to say, the wagon trains that wended their way toward the Rio Grande afforded numerous escape opportunities for any slave teamsters who might have been brought along. But cotton and slaves were not all that crossed the border—a large number of political refugees did as well. Most of the Brazos-area refugees were Germans who had opposed secession and who were now subject to both persecution and, beginning in 1862, a Confederate military draft that they considered unjust. And of course, Brazos residents followed Louis Napoleon's effort to colonize Mexico in the local press. As the war progressed, however, the Brazos moved gradually out of Mexico's orbit. Slavery, which for more than thirty years had kept Mexico in the forefront of black and white Brazos residents' imaginations, began to weaken. To be sure, Mexico continued to attract runaways through the war, but as time passed, Brazos residents, black and white, saw slavery gradually crumble. When it was all over, the boundary between the two nations no longer meant what it had.

Masters and Slaves in Wartime

Brazos society ran according to the paternalistic ideal that envisioned white male heads of household receiving the obedience and deference of their dependents in exchange for protection and sustenance. There were several dimensions to planter paternalism, borderland and otherwise, not the least of which was the ability to furnish material goods, since dependents obviously required sustenance and shelter. But material goods were symbolically significant and were in that sense the lifeblood of antebellum paternalism. Brazos planters practiced paternalism not only by giving their slaves the necessities of life, but also by providing payment and luxury goods as incentives. As long as the goods flowed, the planter could fulfill his obligations and animate his dependent households-within-a-household. Slaves, as we have seen, did not have to internalize the familial metaphor in order for it to work. As long as slaves negotiated in the paternalist idiom, and as long as they could manipulate the metaphor of family to extract concessions, the arrangement survived. But the war interfered with access to credit and markets, throwing the entire system off center.[5]

From the start, the Brazos economy had been nourished by its ties to markets and credit. Since the 1830s, New Orleans had been the financial center for Brazos planters. Crescent City factors provided the credit necessary for the purchase of slaves, seed, land, and machinery, and it was from there that Brazos cotton and sugar went to market. As Texas plantation society grew, Galveston emerged as a regional center, and by 1860 the railroad allowed Houston to compete for the produce of the Brazos and Colorado planters. Still, no matter which city was the focal point, the plantation economy thrived on credit and markets for cotton and sugar. Severing the links between Galveston, New Orleans, New York, and Liverpool could result in serious problems. Without credit or markets, the tenor of plantation life would change. Luxury goods would be scarce. Food crops replaced cotton. Plantation hierarchies would have to be adjusted.[6]

Soon after the advent of the war in April 1861, the United States announced a naval blockade of the southern states. The blockade immediately became the subject of intense diplomatic wrangling between the United States, the Confederacy, and Great Britain, the largest consumer of southern cotton. The Confederate States denounced the blockade, insisting upon their status as a sovereign nation, not subject to the policies of the United States. Confederates also pointed out that their adversaries had failed to provide enough naval muscle to enforce their proclamation and had therefore erected an illegal "paper blockade" unworthy of recognition. The South was also confident that the blockade would not last. Only three years earlier James Henry Hammond had proclaimed the sovereignty of King Cotton. Southerners reasoned that English textile mills depended on American cotton, and it would only be a matter of time before England intervened or, at the very least, called the northern bluff on the blockade. The British, however, were not inclined to interfere. A bumper crop in 1860 had given the mills a surplus, and any shortfalls could be made up by supplies from India and Egypt. In short, the blockade remained in place, ineffectual at first, but by 1863 it effectively cut the Brazos off from Atlantic markets and credit.[7]

Confederate policy had furthermore been to withhold cotton from European markets in an attempt to force British recognition. Texas planters, whose proximity to Mexico gave them outlets beyond the reach of the blockade, successfully lobbied for an exception that would allow them to sell their produce in Mexico. By 1863 Brazos farmers and planters were shipping all of their cotton south to Matamoros. Since no railroad connected South Texas with the cotton-planting areas in East and East Central Texas, much of this was done with massive freight wagons along dirt roads, which meant higher freight costs and long delays in credit. The German community of Cat Spring assembled a caravan and sent seventeen men to escort it to Mexico. The cotton sold for one dollar a pound, but the teamsters earned ten cents of that.[8] Shipment via Mexico exposed cotton producers to other costly problems. In 1862 the news of the sale of John Washington Lockhart's cotton was delayed several months. His Houston factor was certain the first notification had been intercepted by the blockade, and the second notification arrived from Liverpool by way of Tampico, Mexico, 250 miles south of the Texas border. In sum, even though Brazos farmers and planters managed to avoid the fate of cotton producers elsewhere in the South, the uncertainties and costs of shipment through Mexico hurt their profits.[9]

After two years, serious shortages began to plague Brazos society. In normal times, Brazos farms and plantations had been able to provide all or nearly all the necessities of life. Now, with beef, pork, and corn going to feed the army, it was much harder to sustain the household. "If you are forced to come here, let me advise you to bring everthing in the shape of something to eat," Cora Bryan of Independence wrote to her sister-in-law in Brazoria, who was thinking of moving in. Luxury items such as tea and spices could not be had at all. Chicken and eggs were available only at "fabulous prices." Basic items such as butter and flour were "almost as uncertain, as the intervention of France and England." Most importantly, Cora pointed out, procuring everyday necessities was "now reduced to a science, the simple thing of procuring enough to satisfy the absolute gnawings of our imense family," by which she meant, of course, the family's slaves. Unfortunately, records of slave rations and incentive payments do not exist for the war years, but slaves almost certainly felt the effects of wartime shortages. The inability of slave owners to provide their slaves with accustomed material goods had serious implications for a master-slave relationship that relied on the metaphor of family to contain its discontent. The problem was soon felt.[10]

Adding to the economic woes of Brazos planters and farmers was the Confederate policy of impressments, or forced requisitioning of provisions from farms and plantations. The ability of a master to deliver food and other supplies to slaves was a cornerstone of the paternalist posture and an essential currency in master-slave negotiations, and army policy unwittingly played havoc with it. Army patrols marched through rural areas in search of corn, sugar, beef, and pork. The army "impressed" whatever supplies it needed and paid in Confederate notes, which bearers soon learned were accepted at a steep discount when they were accepted at all. From December 1863 through 1864, Stephen Perry's Peach Point plantation received at least five visits from the army. On the first visit, soldiers took five mules, for which Perry was paid $1,875.00. In February troops took seventeen beeves, valued at $595.00. In April he surrendered four cords of wood, worth $400.00, and one hundred head of hogs, worth $1,000.00. Less than a month later, the army impressed more than thirteen thousand pounds of sugar, for which he was promised $3,921.90, and in October he relinquished twenty more beeves, for $1,200.00. Perry's total loss to impressment from December 1863 through October 1864 was at least $8991.90. Given the depreciated state of Confederate notes, he would have had to consider the entire amount a write-off.[11]

Far more detrimental was the policy of impressing slaves. In the vast majority of instances, masters and slaves negotiated in the language of paternalism. In order for the familial metaphor to work, slaves needed to know where final authority rested. Before the war, hired slaves sometimes played owner and hirer off each other, taking advantage of the ambiguity of having to serve two masters. With so many slaves now serving military authorities rather than their masters, it was hard to know whose authority was final. The exclusive authority of a man over his household dependents, virtually sacrosanct before the war, was now compromised. Military impressments of male slaves increasingly weakened the foundations of antebellum slavery, not only among those men who were impressed, but within the slave communities they were forced to leave behind.[12]

Planters wrangled constantly with army officers over the number of slaves impressed. In 1864 Thomas Affleck, owner of 130 slaves, sent an accounting of all his eligible slaves to the enrolling officer of Washington County. There were 15 able-bodied men between the ages of eighteen and forty-five left at his Glenblythe Plantation. He had hired an additional 6 male slaves from John Andrews, bringing his total to 21 candidates for impressment. Affleck did not intend to yield completely to the army's demands, however, and he tried to minimize the number impressed. First he pointed out that his son Dunbar had taken seventeen-year-old Alek with him into the army as a body servant, which Affleck thought should count against his quota. Another slave, William Allen, a skilled molder, was working at a foundry in Hempstead, across the Brazos. Two more slaves, Anderson and Charles Robinson, were working on Buffalo Bayou, near Houston. By Affleck's reckoning, five slaves "is one more than my quota up to the last 6 or 8 weeks; & this independant of the boy in the Army. I claim in return of Anderson. Be kind enough to let me know how I will proceed to procure his return."[13]

Slave owners also worried that army work would harm slaves' health, jeopardizing valuable personal property. Planters often had no idea what their slaves were being used for, how they were fed, or what their general health was. While serving near Galveston, John Washington Lockhart tried to check on his slaves and those belonging to his neighbors back in Chappell Hill. "I cannot hear or see anything of our Negroes," he wrote, "I have looked among several lots of negroes here but cannot find them. I understand that the Government is not issuing any clothes and I expect they are nearly [naked?]. Tell Sallie [his wife] to make up some clothes for Nath and Anderson and if I am so fortunate as to find out where thay are I will let you know."[14]

In addition to their reluctance to lose property valued in the thousands of dollars, slave owners sensed that impressment corroded the reciprocities of the master-slave relationship. First, if army service entailed deprivation of material necessities such as clothing, as Lockhart's letter indicates (and which was a problem endemic in the Confederate armies), impressment undermined the slave owner's commitment to provide for his dependents. The reputation of army camps as filthy, disease-ridden death traps could not have been very comforting, either. Just as seriously, impressment removed the slave from the control of the owner, which deprived the master not only of his labor, but of his discipline as well. Some of the more elitist planters may have worried that their bondsmen would come into prolonged contact with the nonslaveholding whites who made up the bulk of the Confederate armies, who were often seen as vice-prone and dissipated. If we are to take seriously the notion that planters viewed physical punishment as the flipside of benevolence, allowing the army to "correct" refractory slaves would undermine the paternalistic nature of the relationship. The familial metaphor could hardly survive a situation in which the master could neither discipline nor provide for his slaves.[15]

From the slaves' point of view, impressment brought serious disruption to community life. Service could last for months at a time, fragmenting families just as surely as hiring out. Since the army took only men between the ages of eighteen and forty-five, Brazos slave communities would have become increasingly dominated by female-headed households. Just as it is very difficult to know precisely how many antebellum slave households were headed by men, it is equally problematic to estimate how many lost a member and for what lengths of time. Peach Point records showed at least eight distinct slave households in the 1850s and 1860s. Seven had both father and mother present, and one was headed by a woman. In the summer of 1863, the only period for which we have complete records, the army took five men from Peach Point. In June Bill was taken away for a week, and Sam worked two separate seven-day stints. Manes and Arthur were taken into the service from July to October. Then, for three weeks in September and October, Sam and Bill had to return to service, this time accompanied by Houston.[16]

Perry's ledger suggests that he may have pursued a strategy designed to maintain the stability of his plantation and his paternalistic status. His records reveal that he avoided sending male slaves with families to the impressment officer. Two of the slaves impressed that summer, Houston and Sam, were young in 1863, twenty-two and eighteen respectively, and had probably not had time to form families. One slave, "Affrican" Bill, was forty-five at the time of his impressment and appears never to have married. Thus it seems very probable that an effort was made to keep the fathers home, although exactly how Perry and his bondsmen negotiated that arrangement is impossible to say. In the end it seems plausible that Stephen Perry and some of his slaves found common cause in avoiding the disruption brought on by wartime impressment of slaves, but at the expense of the slave community as a whole.[17]

Because documentary evidence is scarce, it is far more difficult to assess the impact of impressment on those slaves who lived on farms rather than plantations, but it would undoubtedly have been more devastating. Demographic evidence alone makes it clear that even in peacetime it was a challenge for small-farm slaves to form stable two-parent families. Now, in time of war, the smaller holdings with fewer adult men could not avoid sending the men away. Forced army service, then, would have had a proportionately more powerful impact on families living on Brazos farms. Lizzie Scott Neblett, whose husband owned ten slaves, noted that the "Negros have been panic stricken about the 'conscripts' as they stile the impressing officer." It is certainly easy to see why they would be. Large plantations could buffer the dislocations by sending out single men; smaller holdings could often do no such thing. The loss of a father, who in some cases would have been the only adult male on the farm, was emotionally painful as well as economically crippling.[18]

As the war progressed, Brazos residents increasingly saw impressment of both produce and slaves as an unreasonable burden and felt justified in evading it outright. Some planters began to hide their goods rather than surrender them for worthless army scrip. Sarah Perry, who left Peach Point to live with relatives in Washington County during her husband's army service, asked her brother to help minimize the plantation's losses. "I took Simon & Billie down & scattered that hogshed of sugar that the Government left so they can not find it anymore," he wrote back, apparently having transferred the one thousand pounds of sugar into smaller containers. "If they ask you about it tell them you do not know anything about it," he said, "It is well taken care of though."[19]

Both slave and free Brazos residents could find common cause in evading the impressment of slaves. When he heard General John Magruder had ordered an increase in impressments, Will Neblett instructed his wife Lizzie to hide the slaves. "I want you to take them to the bottoms if the impressing officers come," he insisted. "If they come while the negroes are in the field send some one to them to tell them." Lizzie, it turned out, was one step ahead of her husband, writing back, "I told the negroes to take the bottom as soon as I heard of Magruder's order. They have not come to our house yet, but the negros told me [troops] were at Yarbons [Yarborough's], & I heard an officer took Hill out of the field."[20]

With the drift from cash to food-crop production, some planters found themselves with more slaves than they needed, prompting an increase in slave hiring, which also added to the instability of the system. Hired slaves were also subject to impressment, and some farmers may have rented laborers to avoid surrendering labor they had bought and paid for. The strategy of a farmer in Washington County further suggests that hiring may have upset gender arrangements within the slave family. When asking for help in hiring slaves, F. G. Brown wrote, "You will pleas see Maj Williams for me and & hire 4 four hands women stout & not Pregnent if they know it," adding, "It is the best to hire women because they are not taken to do goverment work." Brown was not entirely opposed to hiring male slaves, but they should be young enough to be exempt from impressment. "Shoud Maj Williams have boys stout enough to do good Plowing & not subjeck to do goverment work," he said, "I will be willing to take 2 too boys in the plase of too of the women[,] but I prefer to have all stout women if you can get them." Actions such as Brown's only drew attention to the vulnerability of the slaveholders' paternalism. The planters' ability to command obedience was predicated upon their ability to practice paternalism. The blockade deprived them of the material and symbolic means of sustaining their "families," while impressment exposed slaves to work, danger, and discipline other than that of the slave owner. Government service also disrupted slave family life, an integral component of the antebellum order. The principal casualty was paternalism.[21]

A Civil War

In the most peaceful and prosperous of times, the diversity of the Brazos guaranteed that neighbor relations would always be problematic, but they were never so vexed as to bring down the plantation regime. The war, however, was an entirely different context, one that turned minor disputes into major conflicts. Tiffs between planters and overseers, common before the war, took on new meaning amid complaints of class bias in the conduct of the war and a greatly feminized home front. Fears of Yankee invasion and slave rebellion made minor political dissent among immigrants appear to be sedition and violation of racial orthodoxy. The gestures, utterances, and practices that had derived their significance from the antebellum context were inevitably altered as the cumulative weight of the war pressed Brazos society.

The war obviously demanded soldiers, but Confederate policymakers found it difficult to balance military needs with the desire to maintain a stable home front. On the one hand, they needed as many young men as possible under arms to achieve political independence. On the other hand, the plantation regime owed its existence to the continued ability of race and gender to suppress class differences. For the first year of the war, the Confederacy relied on volunteers to fill its armies. By the spring of 1862, as the one-year enlistments expired and new ones slackened, the Confederate Congress approved a draft. In 1863 Congress revised the draft law, this time loading it with exemptions. The one that caused the most controversy was the so-called Twenty Negro Law, which exempted from service one adult male for every twenty slaves on a plantation. The law also allowed draftees to escape service if they could hire a substitute, further alienating many ordinary whites, especially nonslaveholders.

The draft was an affront to antebellum understandings of class, race, and gender. The three categories structured the relationship between planters and plain folk by guaranteeing the equality of white male heads of household. But beyond the slogans usually heard at election time, southern society was inherently hierarchical, and equality was never completely established. Equality appeared real only as long as planters and plain folk could make it work in practice, and that became increasingly difficult. The *Bellville Countryman* invoked racialized images of Oriental despotism when it editorialized, "I know of a number of instances where able bodied men have sought protection under the "nigger" law. . . . Where is the poor man? In the army, fighting like a Turk. . . . The war has made the country quite sickly, and I fear its *constitution* will be impaired, unless some remedy is discovered that will give immediate relief." Conscription was even less popular in the German settlements. Louis Lehmann of Washington County, who was serving in a military unit at the mouth of the Brazos, railed against the material hardships faced by soldiers' families and then told his wife in frustration, "Let the master slaveholders who sit at home do something about it."[22]

The experience of Chappell Hill planter John Washington Lockhart illustrates the growing tension between slaveholders and nonslaveholders. Lockhart was trained as a physician, and although he was not required to join the army, he served as a surgeon at several different posts in Texas. Like many other planters who went to war, Lockhart apparently felt uneasy leaving his plantation under the sole control of his wife, so he hired a man named Wood to help out. Whether Wood was officially engaged as an overseer is unclear. He may have been, but he was still unable to secure an exemption from the draft.[23]

Wood found his tenure at Lockhart's difficult as he tried to run the plantation and discipline the slaves while Sallie Lockhart issued contradictory orders. Lockhart's departure had left the lines of authority on his plantation confused, and Lockhart's slaves exploited the ambiguity with consummate skill. The problems started when a slave woman got sick and "needed something more than I could do for hir." Wood, doubtlessly mindful of the value of a slave, called in a Dr. Perkins to treat her. When Sallie Lockhart found out that Wood had called in a doctor rather than treat the slave himself, she scolded him, saying that Dr. Lockhart had given orders not to summon a physician. Wood resented not having been told of the policy and retorted, "I think I had ort to have be notifide of the fact as I was left with the negrous sometimes 2 or 3 dayes at a time without eneyelce being at hom and have not yet ben told what to do and no mater what acadence was to hapen now I would no what to do." Wood had other complaints that he chose not to mention, but these taken together with the doctor incident "caused me to becom disatfide."[24]

Wood's problems only grew worse after his confrontation with Sallie Lockhart. He noticed that a slave named Charlotte had been picking from fifty to seventy-five pounds less cotton than she was expected to, and Wood "had to whip hir som 3 or 4 times." The next morning, while he was working the cotton press, another slave informed him that Charlotte was not well and could not work, presumably from the beating of the previous day. Wood replied that he did not believe it and told the slave to order Charlotte back to work. At ten o'clock in the morning, Wood went over to the quarter and saw Charlotte's cabin door open. "I took my whip and hit hir about 25 or 30 licks," Wood explained, but "when she told me what was the mater with hir and I stoped." He returned to the press and continued baling cotton, but in the meantime Charlotte went to Mrs. Lockhart, who summoned a doctor. Defensive, and undoubtedly resentful at having been overruled by a woman, Wood remarked in a letter to Dr. Lockhart that "thar was nothing the mater with hir but hir old desease."[25]

Wood found the situation intolerable and exercised the most unchallengeable right he had and the one thing that demonstrated that he was not a slave—he quit. He tendered his resignation to Lockhart in October at the height of cotton-picking time. The Doctor either did not receive the letter or was too furious to answer, because Wood received no response. In the meantime he signed a contract with another planter, this time for a full-fledged overseer's job that carried with it a draft exemption. He closed his letter saying, "I have tride to do what was write and give satfacion to all."[26]

Lockhart reacted with indignation and interpreted Wood's actions as an unwarranted exercise of power over him that put him "in a fix." Upon receiving word of Wood's resignation, Lockhart wrote to his neighbor, Col. William Browning asking for guidance. Wood had "vexed me so much . . . that I hardly new what to say or do," adding, "After doing as much for him as I have done and be treated in that manner, made me feel that I would like to give him a good chokeing."[27]

Of course, such incidents had occurred before the war also. We might even ask whether Wood's conflict with Sallie Lockhart actually demonstrates growing confusion over who was running Brazos plantations, or whether it was just another incident. Considering the context in which it occurred—during a time when wartime demands had greatly feminized the Confederate home front, during a time of instability in the slave quarters, during a time when German Unionists appeared to be an enemy in the Confederate midst—the incident was more than just another plantation squabble. It shows the antebellum modus vivendi, the agreed-upon principles of negotiation among whites of different classes, placed under great strain owing to the war. Clearly, Wood expected to have final day-to-day authority over the running of the Lockhart place; he expected Sallie Lockhart not to undercut his authority over the slaves. Her actions raised a series of questions. What did it mean when the wife of a plantation owner overruled a hired man? Did it mean that white masculinity was under attack? Did it mean that property owners enjoyed privileges not shared by those who did not own land? If this was so, then did race still signify equality as it had before the war created these disruptions?

The meanings of German ethnicity were also altered in the wartime context. The Germans were, on the one hand, white yeoman farmers, in many ways undistinguishable from their Anglo counterparts. On the other hand, German ethnicity struck a dissonant note in Anglo efforts to harmonize race and class. Anglos often suspected—correctly—that many of the Germans harbored antislavery sentiments, although the Germans rarely articulated them publicly. Some immigrants were suspected—again, correctly—of harboring and aiding runaway slaves. Protestant evangelicals among the Anglos also recoiled at the Germans' fondness for drinking wine and dancing on Sundays. The immigrants, for their part, found the nativist impulses of some Anglos to be vaguely threatening, although they found comfort in the knowledge that prohibition and Sunday laws enjoyed little political support.[28]

The politics of secession made it difficult for German identity to continue its liminal existence. Because of its association with Unionism, German ethnicity carried connotations of danger and sedition. Secessionist editors and orators linked Unionism with the specter of slave insurrection, race mixing, and unmitigated evil, vividly captured in the epithet "Black Republican," an insult used hand-in-hand with "Black Dutch." The immigrants had never been enthusiastic supporters of secession, and during the slave-rebellion scare of the summer of 1860, as Anglo Texans searched frantically for abolitionist incendiaries in their midst, suspicion predictably fell on the Brazos Germans. The precipitating incident was the discovery in 1860 that Josef Lesikar and Jan Reymershoffer, two Bohemians living in the German community of Cat Spring, were "agents" for a Czech-language abolitionist newspaper in St. Louis. Anglos were particularly incensed that the masthead of the newspaper carried the names of "Abram Lincoln and Hannibal Hamlin, the 'Black Republican'" presidential ticket, whom they blamed for dispatching incendiaries to the peaceful South. The local newspaper warned that the "circulation of such papers must not be tolerated among us" and in the following weeks charged that a conspiracy between Germans and slaves was afoot. According the editor of the *Bellville Countryman,* on the first Monday in August rebels planned to kill all the men and old women, leaving the young white women to be "parceled out amongst these infamous scoundrels." Anglo residents of Austin County held a public meeting at which they invited the Germans to join them in tightening security and vigilance.[29]

The incident widened latent fault lines in the immigrant community. First, it divided the immigrants by class. Anglo charges of German disloyalty prompted some Germans to prove their fidelity to the Confederacy by denouncing the abolitionists in their midst. The most vocal seem to have been affluent, longtime residents, several of whom owned slaves and had business ties to the Anglo community. They included merchants such as Ferdinand Engelking and Ernst Knolle, as well as the Kleberg and Langhammer families. In fact, it was Knolle who turned in Lesikar and Reymershoffer, the accused abolitionist "agents," and called for the pair to be "hanged from the nearest oak tree." According to Lesikar, while Anglos were merely suspicious, the German slaveholders, led by Knolle, were positively threatening. When a party of six German Confederates paid an unexpected visit to Lesikar, he managed to turn the group away without violence, though he and many others understood themselves to be in great danger.[30]

The second division in the immigrant community was along ethnic lines, as Germans and Bohemians split over the war. Although a clear majority of Germans opposed secession, the Czech-speaking community was nearly unanimous in its opposition. It is difficult to say why, but surviving autobiographical accounts suggest that many of them were Protestants and Czech nationalists, which may have made them sensitive to issues of liberty, conscience, and subordination. They were also, on average, poorer than the Germans, and a number lived in German homes as cooks and hired hands. Thus it is possible that their Unionism was a reflection of class tension directed toward the more affluent Germans, who were also more likely to be pro-Confederate. Charles Nagel, son of a German Unionist who later fled to Mexico, recalled, "From the first the rumor among the children was that the Bohemians were so opposed to the war that they would not enlist; that some of them lived in hiding—even underground—and that those who had been forced to join the army would fire too high to hit the enemy." In many ways, then, class and ethnicity overlapped, with the most affluent Germans on the Confederate side, poorer Bohemians in strong opposition to the war, and the bulk of the German community harboring Unionist sentiments but avoiding overt acts of disloyalty.[31]

Before the war, the Germans had managed to elide the contradiction between ethnic distinctiveness and racial unity by isolating themselves from Anglo plantation society. The isolation was never as real as it seemed, though, and the politics of war forced a dialogue between race and ethnicity. Some Germans served in the major theaters of war in Virginia and the West and counted on being able to return home unmolested when their enlistments expired. Other Germans (and some Bohemians) joined the army or the militia in a gesture of loyalty, hoping to serve near home and planning to desert if sent far away. Jan Ustiník, a Bohemian from Cat Spring, joined an infantry regiment in 1862 and was stationed for a time at Hempstead, only ten miles from home on the other side of the Brazos River. Still, he always intended to desert, and after witnessing part of the battle for Galveston, he fled to Mexico, secreted among the bales of cotton on a wagon driven by his brother. From Mexico he sailed to New Orleans and spent the rest of the war in St. Louis.[32]

Unionism and a resentment of ill treatment by secessionists made the German communities barren territory for Confederate recruiting. An enrolling officer from the Lower Country who was trying to fill his quota by recruiting in the Cat Spring neighborhood wrote that he did not expect to find more than twenty or thirty willing soldiers.[33] An 1863 recruiting drive for the state militia near Cat Spring found only twenty volunteers to fill a company that was supposed to number eighty.[34] Friedrich Winkelmann observed in 1863 that some Germans, in the mistaken belief that only single men could be drafted, were frantically marrying local women. In all of the immigrant settlements, young men hid out to avoid conscription and survived with community sympathy and support. Some lived in abandoned outbuildings and cabins in remote areas. Others lived in the woods or close enough so that they could flee there at the first sign of the provost marshal. Charles Nagel, who lived in the German town of Millheim, near Cat Spring, recalled that his father, a doctor, made house calls, as it were, for Bohemians in hiding.[35]

Relations between Anglos and Germans reached their nadir at the beginning of 1863, with allegations of a seditious plot among Germans in four counties. The "New Ulm Matter," as the press referred to it, is significant not only as a milestone in the alienation of the immigrant community from Anglo society, but for what it reveals about intraethnic conflict over the meaning of Germanness. The plot was first detected just after New Year's Day 1863, the date that Lincoln's Emancipation Proclamation was to take effect, something Anglos saw as a deliberate attempt to spark slave rebellion. The affair also coincided with the brief Federal occupation of Galveston, and some Anglos suspected that the Germans might seize the opportunity to open a second front in Texas. In this atmosphere, Anglos on the Brazos were not only more suspicious than usual of the slave population, but also extra sensitive to the racial and political heterodoxies of the German population.

The events leading up to the New Ulm Matter began in December 1862, when it became clear that the Germans of Austin, Washington, Fayette, Lavaca, and Colorado counties would defy the governor's militia draft order. Early in the month, military authorities began to hear reports of meetings being held in the German settlements, aimed at organizing draft resistance. Then, on December 23, when an enrolling agent near Industry tried to assemble wayward conscripts, a mob drove him out, beating him with sticks and iron bars. With the situation fast spinning out of control, A. J. Bell, the enrolling officer in charge of western Austin County, informed the superintendent of conscripts in Austin that "the Germans of my district and of the adjoining counties are in a state of open rebellion to our Government." Bell went on to report that six hundred immigrants had held a meeting near Shelby (also known as Roeder's Mill) on New Year's Eve and had adopted a resolution organizing the various settlements into companies of infantry and cavalry to man a picket guard. In order to suppress the rebellion, Bell requested "a full regiment of cavalry, to be well mounted, armed, and supplied with subsistence," telling his superiors that anything less "would meet with defeat." Meanwhile, General William Webb, of the Texas State Troops, reported hearing from German and Anglo informants that in addition to resisting the draft, those who attended the meeting (which he estimated at between 500 and 700) had pledged "to stir up insurrection with all of its horrors in case of conflict," referring presumably to the possibility of Union invasion following the seizure of Galveston. Webb was convinced that most of the Germans were loyal, or "true men," but that "the seeds of disaffection have been sown by native Americans." With what he called a "civil war" brewing, Webb echoed Bell's call for cavalry to "overawe the disaffected."[36]

On January 27, a detachment of Confederate soldiers under Lieutenant R. H. Stone, accompanied by a small number of German "guides," arrested six of the alleged ringleaders at New Ulm. Two had addressed the meeting at Roeder's Mill; one was said to be a captain in the clandestine regiment; one was accused of being "an abolitionist in principle" who was reported to have said "that were a runaway negro to come to his house, he would furnish him with a horse and money to effect his final escape"; one was arrested as an accessory for failing to divulge the "conspiracy." One of the detainees reportedly confessed the existence of an organization hostile to the Confederate government that had a large cache of gunpowder. After provost marshals took the prisoners to Columbus for interrogation, several German-Confederate "guides," accompanied by Lieutenant Stone's soldiers, dragged the suspects' families out of their houses and beat them, leaving the wives bleeding on the main road through New Ulm. After being held for a week, the prisoners were released to civilian authorities, who set them free, uncharged.[37]

A petition to military authorities from a group of German women, expressing outrage at violence directed at them on January 27, prompted an investigation by Lieutenant Colonel Henry Webb. The petition apparently alleged that military personnel led the assault on New Ulm, but when Webb solicited statements for his investigation, he found the Germans unwilling to implicate any soldiers from nearby units. Instead, they charged that two men named McElroy, two named Henderson, and a "straggling soldier" had committed the violence, along with unidentified "neighbors." In another statement, taken by Lieutenant William J. Wheeler, five prominent citizens of New Ulm said that the soldiers "acted gentlemanly and with utmost propriety." One of the assaulted women even retracted her original statement to say that while she was knocked down, it was not by any of Lieutenant Stone's men, and that while she had received "one or two scratches from a bayonet," she could not tell whether she "ran against it" or not; and though she was struck in the head with a rifle butt, she now knew that "the blow was not aimed at her" and that she had carelessly run into that, too. The residents of New Ulm had clearly reconsidered the wisdom of pressing their complaint and were now in the process of retracting it in order to deflect the unwanted attention.[38]

The most notable result of the beatings was a very public exchange that took place in the newspapers between Confederate officials along with local Anglos, who charged that there really was a "gunpowder plot" afoot at Roeder's Mill, and an anonymous writer who called himself "Brother Dutch," a self-appointed spokesman for the German community. Here, for the first time, a member of the immigrant community sought to define publicly what it meant to be German in Brazos plantation society. In letters to the region's newspapers, Brother Dutch tried to reconcile German identity with white southern identity without abdicating the right to be German. It was something of a high-wire act: if the outrages committed against the immigrants were excused, the speaker would lose all credence among Germans and Bohemians; but if the spokesman denounced efforts to fight sedition too strongly, he risked being branded disloyal himself. The first part of his strategy was to confirm the loyalty of the immigrant population, which he did by asserting that regimental muster rolls revealed large numbers of foreign men in uniform. He then established the Germans as long-standing, steadfast Texans, "honest men who have resided in Texas from 15 to 25 years, and men who have taken up arms in defence of the country, long before the day of annexation." Finally, Brother Dutch took pains to say that it was the "personal enemies" of the arrestees who beat the women and children, not the Confederate soldiers. To accuse the troops of "leav[ing] the ladies beaten black and blue and senseless in a gorge of blood on the ground" was to condemn the Confederate Government openly and rekindle images of disloyalty.[39] By the time he was finished, Brother Dutch had fashioned an image of Germanness that was marvelously ambiguous, loyal on "southern principles" but decidedly distinct from Anglo identity.

After McElroy and Henderson confirmed their charges in a letter to the *Telegraph* on March 11, Brother Dutch grew more bold in defense, mocking Anglo cowardliness and claiming southern manliness for the German community. In one letter he snickered at the notion that the Germans had a cache of powder hidden near New Ulm. "Have you ascertained how much powder has been concealed about Post Oak Point?" he asked in one letter. "I am fond of smoking, but if there is a danger to blow up this and our neighbor republic—Mexico, I will do without until the inflammable stuff is *removed from the mind* of the people."[40] In another he admonished, "To run down the dutch is a very cheap kind of patriotism," then linked Anglo criticism to womanish gossip by pointing out that "it can be done at home, by the fireside, without much injury to 'OUR' society, and there is no danger of being hurt by attacking peaceable men and innocent women in their night's rest." Then, in a style reminiscent of that most manly of activities, the poker game, he dared McElroy and Henderson to prove their case: "If the ropes can't be used for hanging one set of men, they may be used to transport another. Don't always 'whip the devil around the stump,' but come forward and prove! If the men are traitors, their property, consisting of town lots, farms, stock &c., will be confiscated, and as a matter or course, sold for a song, and in Confederate money, too. There is a speculation, kind neighbors, strike while the iron is hot! Why don't you prosecute? If you are coming, why don't you come along?"[41]

Nobody ever called Brother Dutch's bluff, and his construction of a loyal, white, masculine, yet distinctive German ethnicity was the last word on the matter in the local press. But that Brother Dutch really spoke for the mass of the immigrant community is doubtful. At the same time his letters were published, the local newspapers ran a list of thirty-one army deserters from Austin County, almost all of whom were immigrants.[42] Unknown numbers of Germans and Bohemians fled before ever being mustered into the service. Many of those who stayed behind found themselves swinging from post-oak branches, the victims of unidentified civilian lynching parties.[43] It was painfully obvious to all who cared to see that the immigrant community did not share the same commitment as their Anglo neighbors to southern independence. In effect, Brother Dutch continued the peacetime strategy of harmonizing Germanness with southern racial and political orthodoxy, but it is difficult to imagine his arguments ringing true. The Civil War revealed German identity to be fundamentally at odds with southern Anglo identity, and the dissonance dealt a severe blow to an already-crippled antebellum order.

Millennium

In most of the slave societies in the Americas, rumors of emancipation brought instability. Although slaves had always made efforts to free themselves, it was not until the European Age of Revolution that they sought to destroy the slave system altogether. The reasons probably have much less to do with slaves' sudden achievement of revolutionary consciousness and much more to do with a tactical appropriation of rhetoric and opportunistic political alliances. In any case, it is almost axiomatic that political instability encouraged self-liberation. The classic example, of course, is Saint Domingue, where conflict among whites and mulattos created an opening for the hemisphere's largest and only successful slave rebellion, but there are many others. The 1831–32 Baptist War in Jamaica began when slaves became convinced that Parliament had abolished slavery and that estate owners were suppressing news of "the free." The rumors were not true, but the credence given them by ordinary slaves encouraged a massive resistance that ultimately brought about the same end. In Cuba, during the Ten Years' War of 1868–78, nationalist rebels in the eastern section of the island dealt a serious blow to slavery by declaring emancipation a goal of the rebellion. Even though rebellion was put down, Spanish authorities felt compelled to adopt measures to bring about the end of slavery.[44]

North American slaves were by no means exceptional. Antislavery politics in the early republic may have been a factor in Denmark Vesey's 1822 plot, although links between the 1831 Nat Turner rebellion and abolitionism were purely a figment of white imaginations. In fact, slaves on the Brazos had once before taken advantage of antislavery politics, rebelling as the Mexican army approached in 1835. The promise of emancipation, or even the rumor, was often the catalyst for a testing of the limits of slavery, if not outright rebellion.

It would be a gross exaggeration to suggest that slave resistance in Texas during the Civil War approached anything resembling any of these rebellions, or even one of the lesser unrealized plots. Neither on the Brazos, nor anywhere else in Texas, is there any evidence of an organized effort by slaves to overthrow slavery violently. However, this does not mean that Brazos slaves did not take advantage of the crisis of meaning to test and then redefine the boundaries of slavery. One Brazoria County slave owner spoke of a "runaway mania" caused by the brief Union occupation of Galveston.[45] Indeed, the war had corroded the ideological infrastructure that supported plantation society. Masters could no longer fulfill their own roles as benevolent paternalists once the blockade became a reality and Confederate military impressment took their slaves away. Conscription sapped the ability of race to define equality, while conflicts between planters' wives and overseers did the same for gender. Immigrant Unionism cracked the facade of white unity. Such was the situation on the Brazos when Lincoln announced his Emancipation Proclamation.

Direct evidence is scarce, but undoubtedly slaves on the Brazos knew about the war and Lincoln's proclamation and invested both with revolutionary significance, just as they had done during the Texas Revolution. Beginning in 1860–61, with the secession crisis, the epidemic of alleged arson, and the alleged uprising scheduled for inauguration day, Brazos slaves probably had a more realistic perspective on the meaning of the Civil War than whites in the North or South. True, the incendiaries and insurrectionists may have been imaginary. Yet, as Oscar Addison pointed out, by linking them to the election of Lincoln and injecting them into public discourse, Brazos whites unwittingly helped define the war and thus handed a potent weapon to their own bondsmen.[46]

By 1863, the war had also produced a moral crisis, one that fused with the Emancipation Proclamation to leave the sense that life on the Brazos was approaching cataclysm. In September, the *Bellville Countryman* suggested that defeat and subjugation at the hands of the Yankees represented a sort of divine judgment, asking, "Do we deserve to be Free?" It deplored the evasion of military service by rich planters, "men with subjugation and slavery staring them in the face, occupied in grinding the faces of the poor," then concluded that the South indeed did not deserve to be "free."[47] Rumors of sex between white women and black men represented the crumbling of the last barrier and the ultimate moral transgression for whites. In the Grimes County neighborhood of Lizzie Scott Neblett, gossip abounded that certain white women were sleeping with slaves while their husbands were at war. "The women folks have stirred up quite a muss on Lake Creek, a bout Mrs Magee (the midwife) & [Mrs.] McCane," Lizzie wrote to her husband; "every woman is liable to be talked about, it seems, the state of society is coming to an awful pass." Will Neblett wrote back with news of a black man in Bedias, Grimes County, being lynched for fathering a child by a white woman.[48]

Amid a full-blown crisis of morals, residents responded with a series of revivals. Lizzie Neblett reported a revival in August that had been "raking in the sinners" for a week and showed no signs of stopping. Several of her acquaintances had been saved. The movement carried into the fall, prompting the *Bellville Countryman* to comment, "Altogether, this has been the greatest revival season known in Bellville for many years. The attendance at preaching has been large and attentive and full of good fruits."[49] In fact, the atmosphere on the Brazos in mid-1863 must have seemed positively millennial. Emancipation was now an official northern war aim; planters seemed less paternalistic; whites were no longer equal; a fifth column was at work in the immigrant settlements; white women and black men were having sex with each other; and Confederate forces suffered critical defeats at Gettysburg and Vicksburg.

Slaves would have been well aware of events, and those in Lizzie Neblett's neighborhood seized the opportunity to refashion their relationships with their masters. As we have seen, planters endorsed a subordinate version of paternalistic family relations for their own slaves. Slave masculinity, while in some ways fitting seamlessly into the warp and weft of Brazos society (witness the plots of land at Perry's, which were given to adult male slaves), was fraught with contradiction. Masculinity was a marker of equality in the antebellum world, and the more male slaves were able to force respect for their status as men, the more gender could become a weapon against slavery. In more stable times planters had little to worry about—constructions of race and class were enough to keep slave men from seizing equality on the basis of gender. But as in other areas, antebellum certainties grew ambiguous as the war progressed. Impressment and scarcity had eroded the familial metaphor. The draft had removed most of the young men, and slavery was crumbling in many of the eastern states. In this context it is hardly surprising to find slave men testing the limits of the new environment.

In the late summer and fall of 1863, almost all of these wartime ambiguities were played out on the Neblett farm, near Anderson in Grimes County. Lizzie Scott Neblett's letters to her husband Will chronicle a community in turmoil. Things began to change when Sam, an older, respected slave, began to claim masculine prerogatives. Perhaps the most striking result of Sam's bid for equality was the high degree of neighborhood involvement. The entire community, black and white, watched as Sam challenged his status and quickly became involved in a complex round of negotiations, with future understandings of race, class, and gender hanging in the balance.[50]

Lizzie's husband Will mustered into the army in 1863 and spent most of the war in Galveston. In fact, the war had taken the young, able-bodied men from most of the surrounding farms, leaving several neighborhood farms in the hands of women unaccustomed to slave management. Lizzie was not pleased to see her husband leave, for several reasons, not the least of which was that she felt uncomfortable whipping her slaves, generally seen as a masculine prerogative in the antebellum South. To add to her problems, there was a large contingent of refugee slaves from Louisiana who were hiring around the town of Anderson. Rumor had it that some Louisiana masters had turned their slaves loose in Texas to live off the land and hire in town. These slave owners supposedly got the best of both worlds: they placed their bondsmen out of the reach of Yankee liberators while also relieving themselves of the obligation of providing for them. Such tales were probably untrue, but certainly locals considered refugee slaves a nuisance at best and a potentially serious threat to law and order.[51]

Male slaves sensed the weakness of the old order and began to test the limits of their fast-changing environment. Mrs. Oliver, a neighbor of Lizzie's, sent a neighbor named Langam to whip a slave who "cut up." Another neighbor, "Old Johnny" Driscol, tried to whip a hired slave who "cursed the old man all to pieces and walked off in the woods." The runaway then sent word that he would only return to work in exchange for a promise to refrain from future whippings. Driscol agreed and "never touched him." But Sam, at the Neblett place, made the boldest assertion of manliness. Sometime in August, at the peak of the cotton-picking season, he ran away. "He started to see his wife last Sat after dinner, and returned late monday evening," Lizzie wrote to her husband. And when he returned, Sam began to assert his masculinity in other ways. When ordered to clear a plot of land, he refused outright. When Lizzie sent him on an errand to the sawmill, he spent an entire day away and later claimed to be sharpening the saw. A few days later, when Lizzie got a male relative to whip Sam, he ran away again, crisscrossing the farm twice in order to throw the dogs off the scent.[52]

News of these incidents circulated as neighbors gossiped. "Old Mrs Driscol his wife told Bill Dodd this & he told me," Lizzie wrote to her husband, spreading the story even further, then adding, "Bill says, a great many of the people are actually afraid to whip the negros, I believe him." Slaves would inevitably have heard the stories and passed them on through their own grapevine. Neighborly gossip was the medium through which stories like Sam's were relayed. As the stories made the rounds, the speakers in effect participated in the renegotiation of the master-slave relationship.[53]

In Will's absence, male slaves at the Neblett place redefined slavery quite boldly in a number of symbolically significant acts. Sam, as we have seen, disappeared for three days to visit his wife in Navasota, asserting his fatherly prerogative. Bill and Joe inverted antebellum hierarchy by riding horses and mules rather than walking. Joe somehow acquired his own mule, apparently without Lizzie's approval, and was letting it graze on neighbors' land. Lizzie was aware of this but felt powerless to stop it. "Thornton says Rivers said if I did not make Joe remove his mule, that he would feed all the Gov corn away that he had seen places around the field where Joe had fed his mule," she wrote to Will, adding, "Martin says he will kill Joe if he ever catches him on his place, & several others have been trying to catch him, but he manages to elude them all." More daring was the slaves' use of Will Neblett's fine sheepskin saddle. By riding horses and mules and using the master's saddle, the slaves in effect usurped the prerogatives of their absent master. Despite Lizzie's objections, Bill and Joe did not relinquish the saddle, and they even took steps to hide it in a nearby cornfield. Frustrated with her inability to discipline and control her slaves, Lizzie and several neighbor women prevailed upon a man named Meyers to split time between their farms as an overseer.[54]

Meyers brought drastic changes to the Neblett place. In his first week on the job, he displayed no qualms about punishing refractory slaves, twice whipping Bill, Tom, and Randall at the Nebletts' for "idling away their time" and reportedly "wrought great change" at Mrs. Oliver's as well. The following Saturday he whipped Sam. Sam had been the most assertive slave on the Neblett place, visiting his wife and defying attempts to punish him. The course of neighborhood relations depended on whether or not Meyers could break Sam and put an end to his transgressions.

Meyers soon found out that Sam intended to take to the woods rather than submit to a whipping. Simultaneously, Sam was informed that Lizzie had singled him out as the most troublesome slave at the Nebletts'. But the affair was not that simple. The whipping of Sam was in every way a community affair, with gossip playing a central role. The slave community especially seems to have fanned the flames, perhaps as way to test the limits of the new order they sensed was coming. Lizzie's intense account to her husband captures the slaves' manipulation of the situation, as well as the frantic pace of communication:

> Sam talked to Mrs Olivers Bill, & Bill told Jack, & thus Myers heard the story he told Bill, that Myers should not whip him, no white man should that he would stay in the woods all the year & do no good. . . . Mary with her indiscretion repeated to Sam what she heard me say to Meyers, when he asked if I thought he would have any difficulty in managing any of the negroes. I told him I thought not unless it was Sam, that he might try to run, & not submit to a whipping. When Mary repeated this to him he said to Bob & her your Ma was right I will run, he shant whip me. But Meyers heard the tale at Mrs Olivers & that was enough.[55]

Meyers finally confronted Sam atop a wagon near a cornfield, erecting a worm fence with Randall and Tom, apparently under the direction of Mr. Rivers, a neighbor. Meyers told Sam to come to him, but Sam refused to get down from the wagon, protesting that he had done "nothing to be whipped for." Meyers then threatened to knock Sam off the wagon, which prompted Sam to get down, then start running away. Meyers fired a pistol, which was loaded with shot, but missed as Sam disappeared down a turnrow. At that point, Meyers and two slaves, Randall and Tom, took off in pursuit, catching up to him in the field. Sam, wielding a fence stake, fended off the three pursuers, "tell[ing] them if they layed hands on him he would kill them." Meyers, apparently having dropped his gun, tried to subdue Sam with the butt of his whip but could not get close enough. The two faced off until Meyers called upon Rivers to bring his gun, telling Sam that he would be shot if he did not surrender. Sam gave in, and soon after Meyers whipped him. "Give him bad whipping too," Lizzie related to her husband, "worse than he intended he admited after he came to the house."[56]

For several days, Sam was wracked with pain from the beating. He complained of sore, but not broken ribs, "colic symptoms," chills and fever, all of which apparently frightened Lizzie enough for her to call Dr. Foster, who gave Sam some pills and ordered him to stop drinking water. Lizzie felt guilty for her role in the affair but wrote Will that "tho' I pity the poor wretch I don't want him to know."[57]

But while Sam lay suffering from the beating, the slaves at Nebletts' and neighboring farms continued to manipulate the situation through gossip. They seized the initiative when Mr. Rivers, the neighbor who was in the field with Meyers and Sam when the incident began, apparently let on to a Mr. Hunt's slaves that it was a "damned shame the way Sam was whipped." Hunt's slaves then relayed that comment to Lizzie and insisted that she talk about it with Rivers. Lizzie had little inclination to speak with Rivers about the incident and in fact never sent for him. But she soon found out through the grapevine that Jack Oliver was at the Hursts' and heard a slave tell Rivers that Lizzie wanted to see him. Then Kate, another slave, told Lizzie that Mr. Rivers wanted to speak with her and added that what happened to Sam "was a d—m shame." Lizzie, whose integrity was now at stake among both blacks and whites, told Kate to "hold her mouth, that I did not want Rivers to tell me anything [and] that I know Myers would not have whipped him if he had not deserved it & that it was a wonder he left live in him."[58]

But Rivers's criticism of Meyers's brutality had now become an issue among whites. The Nebletts' neighbors were acutely aware of the slaves' ability to exploit disagreements among whites and understood the danger in Rivers's defection. Dr. Foster, who had visited Sam a few times, thought Rivers had "acted very wrong to talk in the manner he did" and that Sam was still as defiant as ever. But the slaves had now seized the advantage, as Rivers's criticism focused attention on Meyers. In fact, Rivers began to offer a different interpretation of Sam's initial vow that no white man would whip him. Now Rivers thought Sam might have said that he meant Meyers in particular, not white men in general. By this time Sam was well enough to contribute to the communal interpretation of the incident. He now implied that Meyers was a bad overseer who made unreasonable demands on the slaves, maintaining that Meyers had attacked him when he told him that "no body could please him no way." The story now hinged on Meyers's deportment, and he apparently felt uncomfortable enough to admit that he had lost his temper and had been too severe. In effect, the slave community was able to win a tactical victory on Sam's behalf. With Meyers's management style now the issue, Sam was safe from charges of overt rebellion.[59]

The flow of information, gossip, and interpretation did not all break in Sam's favor. Little did he know that Sarah, a slave who was particularly close to Lizzie, supplied her with intelligence gained by eavesdropping on his conversations with other members of the slave community. While Sam was staying at the house of a slave named Nancy, Sarah overheard him say that he never again intended to work under Meyers and would hide in the woods as long as he was there. Sam then added that he laid the blame for his whipping at Lizzie's feet, not Meyers's, showing that the community's success at defining the incident as a matter of bad slave-management did hold for him. Major, in whom Sam confided, then advised Sam to approach Lizzie and tell her that he could no longer work under Meyers; in short, he counseled invoking paternalism in order to restore stability. It apparently worked. We know of no more conflict at the Neblett place. But it was in all probability merely a tactical retreat. As the news of military defeat trickled in, and as impressment, conscription, flight as refugees, draft evasion, and shortages all continued apace, Brazos slaves knew that time was on their side.[60]

When General Gordon Granger announced emancipation in Galveston on June 19, 1865, he did so in an environment already changed by four years of war. The Civil War along the Brazos was a social revolution, but not in the model of Haiti or even of other parts of the South, where military occupation was a much greater factor. Change came as a result of everyday actions by ordinary people in mundane face-to-face situations. What was unusual was the context, which lent new meaning to ordinary actions. From this perspective, no one thing—the impressment of slaves, arguments among whites, or the events at Lizzie Neblett's farm—was "radical" in and of itself. That they occurred together, simultaneously, is what made them radical. Wartime shortages, class conflict, ethnic dissonance, and slave resistance all played off one another to bring change to the Brazos. And though nobody commented on it at the time, war and the crumbling of the slave system initiated a reorientation for Brazos residents. With slavery seemingly doomed, the differences between the Brazos and Mexico were growing less significant. Soon, the meaning of the Texas-Mexico border for those who lived along the Brazos became completely transformed, and soon all eyes were focused northward, rather than southward.

Epilogue

BORDERED LANDS

The Civil War and emancipation marked the end of an era on the Brazos. For four decades Brazos plantation society had felt the gravitational pull of two separate republics, each born in the Age of Revolution, each heir to the era's antislavery imperative. But in the 1820s, with independence secured and sovereignty established, the two republics approached the key question of slavery very differently: one was facilitating its expansion, the other its abolition. The tension between these two ideals was one of the most important contributors to the Texas Revolution and the founding of the slaveholding Texas Republic in 1836. Yet while the defeat of Mexico and the redrawing of the boundaries of sovereign power certainly diminished the influence of Mexico over the Brazos, they did not eliminate it entirely. Mexico never acknowledged the legitimacy of the Texas Republic and still hoped to reestablish sovereignty over the region. The Brazos was still part of a contested borderland. Even annexation by the United States in 1845 did not eliminate Mexican influence. Mexican immigration and land policies had set the fundamental migration, demographic, and property-holding patterns for the region. Mexican law carried over in several important areas, including land and marriage. And, perhaps most importantly, Mexico persisted as a symbol of antislavery long after its claims to sovereignty over Texas were abrogated at Guadalupe Hidalgo. Thus, even after 1848, Texas and the Brazos remained part of a contested, hybridized borderland, despite the treaty. Then in 1861, Anglo-Texans made the fateful decision to secede once more from a republic that seemed a threat to the slaveholding interest. The miscalculation brought about, four years later, the end of slavery and revolutionary change to the region's labor system. The region's status vis-à-vis the border was also altered as a result.

It had been a curious thing, this plantation society in Central Texas. There was certainly nothing like it in the rest of Mexico. In most important respects, its development was similar to that of the cotton-planting regions of the southern United States. The region was settled primarily by Anglo American Protestants and African American slaves and had only a very small population of Spanish-speaking, Catholic Mexicans. Instead of intermarrying with the indigenous residents, Brazos settlers followed the example of the other cotton regions and engaged in their own version of Indian removal, expelling the Karankawa people in the 1820s and the Tonkawas in the 1850s. The population grew, fueled by migration from the east and slave importations. Cane eventually supplemented cotton-growing in the Lower Country. By the late antebellum period, the composite portrait as sketched by travelers was of a decidedly southern place, as indeed it was in many respects.

But there were differences as well, which were largely traceable to the region's borderland situation. Slavery, the most important economic institution in Texas, was deeply affected by the policies and presence of Mexico. For most of his tenure as empresario, Stephen F. Austin had seen slavery as essential to his enterprise, and his colonists emphatically agreed. The legality of slavery in Austin's Colony, however, was complicated by conflicts between Centralists and Federalists over the relative powers of the national and state governments. So while the antislavery impulse of the Mexican and Coahuila-Texas governments cannot seriously be doubted, the confusing torrent of statutes, constitutional provisions, colonization laws, executive orders, exceptions, and "clarifications" issued at the various levels of government gave Anglo settlers the legal ambiguity they needed to begin transforming the Brazos into a slave society. But their success came at a price: bypassing Mexican objections to slavery only weakened the institution by creating a powerful antislavery rallying symbol in Mexico itself. Anglo resistance to Mexican antislavery efforts gave focus and strength to slave resistance. The Revolution of 1836 and the drawing of a boundary between the free Mexican republic and the slaveholding Texas Republic gave the tension between slavery and freedom a geographic tangibility it had never before had. Annexation by the United States in 1845 did little to alter that situation. Thousands of slaves, with "Mexico in their Heads," fled southward. Masters were forced to take this context into consideration in making their purchases and in negotiating with their slaves. They had managed to secede from Mexico, but they remained in a borderland.

Mexican law had left another key institution, marriage, on an uncertain foundation. Significant in its own right, marriage had taken on added importance in the Slave South, as ministers, politicians, and proslavery ideologues analogized it to that other "domestic institution," slavery. Once again, Mexican law did not meet the needs of Anglo slave society. Insofar as slavery was legal in Mexican Texas, the older Spanish codes that had governed the institution gave slaves marriage and family rights. These laws may have been a dead letter in practice, but they served as a reminder that slave societies did not cross borders without undergoing adjustment. Colonization statutes and land laws encouraged the formation of unorthodox "families," at times allowing men to group together as if married. And finally, the practice of bond marriage during the Mexican era allowed for easy divorce, something the rest of the Slave South was loath to adopt. The ideological melding of slaveholding and marriage, which allowed planters and yeomen to join together on the basis of shared mastery, was incomplete in Texas, where some domestic dependents—that is, married women—exercised the right to sever their ostensibly permanent and organic relationship with the head of the household. Even after independence, legislators found it expedient to leave many of these features intact.

Mexican influence on the Brazos manifested itself in still other ways. The population, which at first glance seemed to mirror that of the Lower South, was actually more diverse than that of any other plantation region, with the possible exception of Louisiana. The Anglo and African American population was augmented by a significant number of Germans, whose migration had been welcomed by the Mexican government partly as a counterbalance to growing fears of Anglo secessionism. The Germans established themselves in several enclaves in the Upper Country, their numbers growing through chain migration. With few exceptions, the Germans were not overtly hostile to slavery or to their slaveholding neighbors. They held a range of views on the question of bondage; some acquired slaves and others silently cursed the institution's evils. The majority probably occupied a middle position, in which they found chattel slavery a vaguely alien and troubling concept, but their feelings were not strong enough to risk the hostility of their Anglo neighbors by advocating antislavery policies. In any event, a steady stream of new immigrants supplied the labor needs of the German settlements. As the politics of slavery in the United States grew more acrimonious in the 1850s, it became harder to remain silent and aloof. The Germans' public silence on the matter did not go unnoticed, nor did the fact that only a small proportion of the Germans ever became slaveholders. As a result, the Brazos Germans were perceived as being hostile to slavery, regardless of their true sentiments. A curious sort of nativism developed, a fusion of the proslavery and anti-immigrant movements, giving secession and wartime politics a distinct ethnic dimension. Such was the legacy of Mexican immigration policy.

But now, in 1865, slavery was dead. Since the 1820s, Mexican hostility to the institution had been a dissonant force in Anglo efforts to create a southern-style plantation society. Since the 1830s the U.S.-Mexico border had been to a large extent defined by the existence of slavery on one side and its absence on the other. With slavery suddenly gone, the meaning of that border changed: it no longer embodied the dream of freedom to a majority of Brazos residents. It still invited occasional strategizing. In the wake of emancipation, some Washington County planters removed to Mexico, where they established plantations based on wage labor. Not satisfied with the results, most returned within a few years. Former slaves sometimes crossed the border as well. Most of these were individuals searching for employment or a less-repressive racial environment, or perhaps to escape the law, but in 1889 a black-led scheme to establish a settlement in Mexico attracted interest from seven hundred families from Washington, Brazoria, Wharton, and Bastrop counties.[1] For the most part, though, Mexico receded in the consciousness of Brazos residents, black and white. For African Americans in particular, the U.S. government replaced Mexico as a symbol of freedom and equality. In the years after emancipation, the Brazos became a truly bordered land.

Emancipation was what altered the significance of the border, and its impact on virtually every other phase of life along the Brazos cannot be overstated. After eroding slowly during the Civil War, slavery ended quite suddenly when General Gordon Granger landed at Galveston on June 19, 1865, and decreed, "All slaves are free." But then, setting a pattern for what followed, he immediately advanced a definition of what it meant to be free: "This involves an absolute equality of personal rights and rights of property, between former masters and slaves, and the connection heretofore existing between them, becomes that between employer and hired labor." He went on to advise former slaves to "remain at their present homes, and work for wages" and warned them against "idleness." His understanding of freedom was consistent with the ideals of the free-labor society that he represented, with policies well honed during the four-year destruction of slavery elsewhere in the South. It was not, however, the only possible definition of freedom. The erstwhile slaves of the

Brazos had an entirely different conception, one that envisioned independent black communities underwritten by the ownership of land, a view that stood in sharp opposition to the policy of Granger and federal officials. Brazos planters—transformed from "labor lords" into landlords by emancipation—advanced a third distinct definition of black freedom. With no other realistic choice, they accepted the demise of slavery, but with an eye toward securing labor for their lands, they defined freedom simply as an end to human chattel and the beginning of waged labor. Beyond that, the labor system they envisioned was quite repressive, with a legal code that granted employers wide latitude to regulate laborers, coupled with vagrancy and apprenticeship laws that gave freed people little choice but to sign long-term labor contracts.[2]

At the outset, there was a harmony of interest between landowners and the federal government. Neither party wanted former slaves to leave the plantations, so government agencies like the Freedmen's Bureau worked to create a wage labor force for the plantations. By January 1866, Edward Gregory, assistant commissioner of the Freedmen's Bureau, estimated that 90 percent of the former slaves on the Brazos were under contract to a landowner.[3] Yet even as this achievement was announced, southern landowners and federal officials were falling out on the question of the meaning of freedom. These differences led in 1867 to the imposition of military rule under the Reconstruction Acts and the ratification of the Fourteenth Amendment to the Constitution the following year. The meaning of freedom as advanced by federal officials was largely negative, in the sense that it guaranteed freedom from certain things: from physical coercion, fraud, and openly racist and exploitative laws. But it guaranteed little in a positive sense, neither land nor employment. As most federal policymakers saw it, this absence of positive guarantees was the very thing that made the wage system work—why else would someone perform plantation labor if not for the fear of starvation? When the Freedmen's Bureau closed down in 1869, even these negative guarantees were left on a shaky foundation.[4]

What evolved, in the meantime, was a hybrid system of land tenure and labor, the result of a bitter, often violent negotiation among the three interested parties. Often called the "sharecropping system," the new system was in fact a great deal more complicated than the name implies. As elsewhere in the South, the Brazos developed a system of tenancy, in which some workers, defined as "tenants," paid rent in cash or in a share of the crop. Others, legally defined as "croppers," were compensated in a share of the produce. Casual wage labor was a third category, in which a worker was simply compensated by the day or month without being responsible for any particular piece of land. This system was hardly unique to the Brazos. Most parts of the South saw similar developments. The one unusual feature of the Brazos was the emergence of a class of smallholders in Brazoria County, where a postwar decline in sugar planting allowed some freed people—perhaps 15 percent of black households—to purchase very small tracts of land, usually fewer than twenty acres. These plots were not enough to guarantee the independence they had sought in the wake of emancipation, but they did permit some latitude in deciding when and how to work on the plantations.[5]

Political life was reconfigured as well. In the immediate aftermath of the war, President Andrew Johnson allowed the formerly seceded states to frame new constitutions and apply for readmission to the Union. The terms were relatively forgiving of one-time secessionists, and they were designed to make the process of "Restoration" a swift one. Texas's government, framed and controlled largely by former slaveholders, was like most, passing labor legislation favorable to landowners and excluding black residents from citizenship. After much repression and violence was perpetrated against the freed people, federal officials dismantled the Johnson governments and set about constructing a new body politic, rooted in universal male suffrage. The Republican Party was the vehicle of black political participation. It also took in a number of former Unionists, particularly Germans. The Republican Party controlled the state until "redemption" in 1873, at which point Democrats, more attentive to landowner needs, came to power.[6]

Despite losing power at the state level, the Republican Party remained viable along the Brazos for several years, sustained by the vast majority of African American voters and a substantial number of Germans. Here was an interesting echo of the antebellum period, with the Anglo nightmare of a German and African American conspiracy realized as never before. But this coalition of convenience—alliance would be too strong a word—crumbled in the real world of New South politics. Simply put, black and German Republicans became Republicans for different reasons. African Americans advocated a Radical Republican iteration that emphasized civil rights and labor issues such as a more just lien law. Germans envisioned a more conservative party of Unionists, largely in reaction to wartime abuses at the hands of Anglo secessionists. As early as 1869, the Germans in Austin County seem to have shifted their support to the Democrats. Many had been in Texas for a long time, had accumulated handsome amounts of property, and shared the same views as Anglo landowners regarding black labor. In places like Industry, there was now a generation of Texas-born German-speakers whose politics and racial views more closely resembled those of the Anglo Democrats. Things were different in Washington County, however. The immigrant community there consisted of more recent arrivals, who, if they were not quite racial egalitarians, were less receptive to the Democratic message. A number of them had had unhappy experiences working on Anglo-owned plantations after the war, and those no doubt played a role in their opposition. In 1873, the Republican-controlled state government divided Austin County into two parts. The portion west of the Brazos was envisioned as a Democratic stronghold, with a white majority dominated by Germans. The section east the river, now Waller County, was meant to be a Republican bastion, sustained by a black voting majority.[7]

Republicans dominated local politics along the Brazos (with the exception of Austin County) well into the 1880s. Then in 1884, Washington County Democrats—billing themselves as a nonpartisan and reform-minded "People's Ticket"—staged a violent coup d'état, attacking African American polling places and destroying the ballot boxes. One of their targets was Chappell Hill, where two black election workers were wounded and one was killed. The county was "redeemed," but Republicans did not surrender easily. They organized and campaigned vigorously during the 1886 canvass. A new German newspaper was founded to support the Republican ticket, counterbalancing the established Democratic-leaning one. After a heated campaign, Democrats once more attempted to steal ballot boxes on election night. This time, a group of black Republicans at one polling place met the thieves with gunfire, killing the son of a local landowner. The incident prompted local whites to lynch three black Republicans and run several German Republicans out of the county, effectively ending all opposition. Acquitted in a farcical trial in Austin, the leaders of the "Washington County Outrage" were forced to testify before the Senate Committee on Privileges and Elections, which was then contemplating a reinstatement of federal supervision of southern elections. Although the testimony was damning, no bill was ever passed. The so-called Force Bill died in a filibuster, supported, of course, by the Texas delegation.[8]

Within a few years of the Washington County Outrage, Democrats overthrew the Republican majorities that governed Fort Bend and Brazoria counties. In Fort Bend, a group known as the Jaybirds confronted the local Republican establishment, known as the Woodpeckers. When a Jaybird leader was shot and killed in 1889, a white mob drove most of the black Republican leadership out of the county. Republicans held on, however, until a series of attacks on Democrats resulted in a wild shootout in the town square of Richmond. At that point, Democratic governor Sul Ross interceded and brokered a "deal" in which the Woodpeckers ceded power to the Jaybirds. Next door in Brazoria County, Democrats took until 1894 to seize control. Organizing as the Tax-Payer's Union, they issued a none-too-subtle threat to any African American who might be inclined to take elected office, thus accomplishing the same end as in Fort Bend, without the same amount of bloodshed. Grimes County held out the longest. There, the biracial People's, or Populist Party enjoyed significant support. That ended in 1901, in a flurry of race baiting, culminating in the wounding of the white Populist county sheriff.[9]

By the turn of the century, it was easy to forget that the Brazos had once been part of Mexico. It had all the features of New South society: a white, landowning elite; a black laboring class, mostly landless; an economy still dominated by plantation crops, particularly cotton; a white-supremacist Democratic Party that dominated regional politics; a crop lien; and periodic lynchings. With the coming of the new century, the Brazos added cattle ranching, prison and corporate farms, a sulfur works, and—this being Texas—oil.

Yet even then, the region's borderland past was still visible if you knew where to look. German was still widely spoken in the Upper Country, although hyperpatriotism during World War I put German-speakers on the defensive, and the language began a slow decline that lasted for much of the century. And German was not the only linguistic vestige of the Mexican era. A few Africans survived into the twentieth century. As late as 1913, ninety-year-old Ned Thompson still "remember[ed] the customs and habits of the wild people of his native land. . . . Sp[oke] the language or dialect of his tribe fluently. . . [and] Retain[ed] a vivid recollection of his early life."[10]

Another reminder of the region's past arrived in the form of migration from Mexico. The trend began early in the twentieth century, as landowners began to hire Mexican laborers to work on cotton farms. Most migrants went to the blackland prairie areas to the west and north of the Brazos, but a portion came to the cotton "ranches" of the Brazos.[11] The trend continues to this day. In predominately rural Austin, Waller, and Grimes counties, the

Hispanic percentage of the population is currently between 16 and 19 percent. In recent years, the one-time rural counties of Fort Bend and Brazoria have become home to several exurban bedroom communities of Houston. The growth has attracted Mexican immigrants, who work in restaurant, industrial, landscaping, construction, and child care jobs, as well as in white-collar positions. Brazoria County's population has doubled since 1980, with the Hispanic proportion increasing from 12.4 percent in 1982 to nearly 23 percent in 2000. As the Spanish-language signs announce, numerous local businesses cater to the county's immigrant population.[12]

Anglo residents tend to view this migration as a new phenomenon, but the continuities with the past stand out as much as the changes. After all, it was Mexican border and immigration policy that led to Anglo, African American, German, and African migration into the region in the first place. And nineteenth-century Mexican land policy still makes itself felt today, as residents continue to litigate over the boundaries and legal status of old claims, often with an eye toward subsoil rights. (On several occasions while I paged through deed books in county courthouses, residents asked guardedly whether I was an oil or gas man and were relieved to find out I was "only" a historian.) In this environment, the border has, not surprisingly, begun to command some of the attention it did in former days. A national movement is now afoot to "strengthen" the border, and paramilitary vigilantes now patrol it regularly. Movements like this are generally blind to historical irony. In early 2006, when Houston was to receive its first Major League Soccer team, a controversy erupted over the name. In a poll, fans chose the name "Houston 1836," following the European tradition of placing the date of the club's founding after its name—though here the date referred not to the founding of the club but to the founding of the city and, not coincidentally, to independence from Mexico. Sensitive to the triumphalist connotations of the name and aware that a significant part of the team's fan base would probably be Mexican, the team changed its name to the "Houston Dynamo." Among the critics of the change was a Brazoria County resident who wrote a letter to the editor of the *Brazosport Facts*, the local paper. "It seems the new Houston soccer team has finally come up with a name that isn't supposed to offend anyone," he wrote. "Just think, if we'd been this politically correct in 1836, the Mexican border would be Oklahoma." Mexico was once again in Brazos residents' heads; the past was not yet past.[13]

Abbreviations Used in the Notes and the Appendixes

ACC	Austin County Clerk, Bellville, Texas
AP	Eugene C. Barker, ed., *The Austin Papers*
BCC	Brazoria County Clerk, Angleton, Texas
BCHM	Brazoria County Historical Museum, Angleton, Texas
DBCAH	Dolph Briscoe Center for American History, University of Texas
DCAC	District Court of Austin County, Bellville, Texas
DCBC	District Court of Brazoria County, Angleton, Texas
DCWC	District Court of Washington County, Brenham, Texas
GLO	General Land Office, Austin
HML	Hill Memorial Library, Louisiana State University
LOC	Library of Congress
LT	H. P. N. Gammel, comp., *The Laws of Texas*
OR	U.S. War Department, *The War of the Rebellion: A Compilation of the Official Records of the Union and Confederate Armies*
RL	Rosenberg Library, Galveston, Texas
SHC	Southern History Collection, University of North Carolina
SHQ	*Southwestern Historical Quarterly*
TSHAQ	*Texas State Historical Association Quarterly*
TSL	Texas State Library
WCC	Washington County Clerk, Brenham, Texas

Appendix 1

TABLES

TABLE 1. Slave and free population, 1850–1860

	1850				1860			
County	Slaves	% Slave	Free	% Free	Slaves	% Slave	Free	% Free
Austin	1,549	40.3	2,292	59.7	3,914	38.6	6,225	61.4
Brazoria	3,507	72.4	1,334	27.6	5,110	74.9	2,033	29.8
Fort Bend	1,554	61.4	979	38.6	4,127	67.2	2,016	32.8
Grimes	1,680	41.9	2,328	58.1	5,468	53.1	4,839	46.9
Washington	2,817	47.1	3,166	52.9	7,941	52.2	7,274	47.8
Total	11,107	52.4	10,099	47.6	26,560	54.6	22,387	46.0

SOURCES: U.S. Census Bureau, *Statistical View of the United States;* U.S. Census Bureau, *Population of the United States in 1860.*

TABLE 2. Landholdings, Washington County, 1837–1860

Size in acres	1837	1845	1850	1860
1–100	5	33	62	162
101–500	17	12	193	314
501–1,000	12	39	57	97
1,001–2,000	25	47	39	68
2,001–4,000	32	32	11	18
4,001–8,000	38	25	4	3
8,001+	16	9	0	1

SOURCES: Texas County Tax Rolls, Washington County (microfilm), pt. 1 (series 1, roll 637), TSL.

TABLE 3. Slaveholding among persistent planters, 1850–1860

Community	Persistent slave owners	Slaves in 1850	Slaves in 1860	% Increase
Chappell Hill	9	148	355	140
Gulf Prairie	6	536	748	40

SOURCES: U.S. Census Bureau, Free Schedules, Washington County, Texas, 1850, record M432, roll 916; 1860, record M653, roll 1307; Brazoria County, Texas, 1850, record M432, roll 908; 1860, record M653, roll 1289; Slave Schedules, Washington County, 1850, record M432, roll 918; 1860, record M653, roll 1312; Brazoria County, 1850, record M432, roll 917; 1860, record M653, roll 1309 (microfilm).

TABLE 4. Foreign born, 1850–1860

	1850		1860	
County	Number	% of whites	Number	% of whites
Austin	810	35.4	1,205	24.0
Brazoria	173	13.0	73	4.4
Fort Bend	53	5.4	193	6.0
Grimes	143	6.1	275	6.0
Washington	117	3.7	1,251	20.8

SOURCES: U.S. Census Bureau, *Statistical View of the United States*, 309, 315; U.S. Census Bureau, *Population of the United States in 1860*, 487–489.

TABLE 5. Known slaving voyages to Texas, 1833–1837

Name	Date	Origin	Number	Destination
Leander McNeel	pre-1835	Cuba	unknown	Brazos
McNeel	pre-1835	Cuba	unknown	Brazos
Ben Fort Smith	pre-1835	Cuba	unknown	San Bernard
Thomas League?	1834	Windward Coast, Slave Coast?	unknown	unknown
Thomas League?	1835	Windward Coast, Slave Coast?	unknown	unknown
James Fannin	1835?	Cuba	152	San Bernard
Monroe Edwards?	1835?	Cuba	183	Brazos
Monroe Edwards	1836	Cuba	170	Brazos
Sterling McNeel	1836	Cuba	40?	San Bernard
Moro and Coigly	1836	Cuba	200	Sabine
unknown	1837	Cuba	<40	Brazos
unknown	1837	Cuba	<40	Brazos

SOURCES: William P. Zuber to Eugene C. Barker, March 12, 1902, Barker Papers, DBCAH; David Randon, Plaintiff in *Error v. Thomas Toby*, in Howard, *Reports of Cases Argued*, 11:500; Robbins, "The Origins and Development of the African Slave Trade into Texas, 1816–1860"; William S. Fisher to S. R. Fisher, March 2, 1836, Convention Records, Archives Division, TSL; Adams, "British Correspondence concerning Texas," 199–200.

TABLE 6. Household structure in two neighborhoods, 1850–1860

Type	Gulf Prairie, 1850 No.	Gulf Prairie, 1850 %	Gulf Prairie, 1860 No.	Gulf Prairie, 1860 %
Nuclear	5	20.8	10	31.3
Nuclear with nonfamily	10	41.7	5	5.6
Extended	0	0.0	3	9.4
Extended with nonfamily	1	4.2	1	3.1
Solitaire	2	8.3	3	9.4
Nonfamily	5	20.8	5	5.6
Ambiguous	1	4.2	5	5.6
Total	24		32	

Type	Chappell Hill, 1850 No.	Chappell Hill, 1850 %	Chappell Hill, 1860 No.	Chappell Hill, 1860 %
Nuclear	35	62.5	89	54.3
Nuclear with nonfamily	5	8.9	41	25.0
Extended	4	7.1	10	6.1
Extended with nonfamily	0	0.0	2	1.2
Solitaire	5	8.9	5	3.0
Nonfamily	4	7.1	1	0.6
Ambiguous	3	5.4	16	9.8
Total	56		164	

SOURCES: U.S. Census Bureau, Free Schedules, Brazoria County, 1850, record M432, roll 908; 1860, record M653, roll 1289; Washington County, 1850, record M432, roll 916; 1860, record M653, roll 1307 (microfilm).

TABLE 7. Washington County divorce cases, 1839–1857

Plaintiff	Abandonment[a]				Cruelty[b]			
	Div	Deny	Drop	Unkn	Div	Deny	Drop	Unkn
Husband	4	2	1	0	0	0	1	0
Wife	11[c]	1	2	0	4	0	4	0
Unknown	0	0	0	0	0	0	0	0
Total	15	3[d]	3	0	4	0	5	0

Plaintiff	Adultery[e]				Unknown				Totals
	Div	Deny	Drop	Unkn	Div	Deny	Drop	Unkn	
Husband	2[f]	0	0	0	1	1	2	0	14
Wife	0	0	0	0	1	0	1	1	25
Unknown	0	0	0	0	0	0	2	1	3
Total	2	0	0	0	2	1	5	2	42

SOURCES: District Court of Washington County, Final Record, books A–R; District Court of Washington County, Minutes, books A–E, District Clerk of Washington County, Brenham, Texas.

[a]Includes cases in which charges such as adultery and cruelty were made in addition to abandonment.

[b]Includes charges of adultery if emphasis was on cruelty.

[c]Includes four cases in which cause is unknown but assumed to be abandonment because defendant did not appear in court.

[d]Includes cases dismissed by judge for lack of jurisdiction.

[e]Includes charges of cruelty if emphasis was on adultery.

[f]Wife had not divorced previous husband. See Final Record, book C, 187.

TABLE 8. Household structure, Cat Spring, 1850–1860

	Cat Spring, 1850		Cat Spring, 1860	
Type	No.	%	No.	%
Nuclear	24	49.0	88	75.9
Nuclear with nonfamily	4	8.2	3	2.6
Extended	3	6.1	15	12.9
Extended with nonfamily	0	0.0	0	0.0
Solitaire	1	2.0	0	0.0
Nonfamily	0	0.0	0	0.0
Ambiguous	17	34.7	10	8.6
Total	49		116	

SOURCE: U.S. Census Bureau, Manuscript Census, Austin County, 1850, 1860, microfilm ed.

TABLE 9. Corn, cotton, and sugar ratios, 1849–1859

	Corn (bushels)/cotton (400 lb. bales) or corn/sugar (1,000 lb. hogsheads)	
County or *Neighborhood*	1849	1859
Austin County	46.6	not available
Cat Spring	68.0	31.6
Brazoria County	28.9*	19.6*
Gulf Prairie	20.0*	21.1*
Fort Bend County	53.1*	16.4*
Grimes County	60.7	20.5
Washington County	40.4	23.3
Chappell Hill	54.2	18.8

SOURCE: De Bow, *Statistical View of the United States;* U.S. Census Bureau, *Agriculture of the United States;* U.S. Census Bureau, Agricultural Schedules, Austin County, Texas, 1850, record T1134, roll 2; 1860, record T1134, roll 3; Brazoria County, 1850, record T1134, roll 2; 1860, record T1134, roll 3; Washington County, 1850, record T1134, roll 3; 1860, record 1134, roll 7 (microfilm).

*Since Brazoria and Fort Bend counties produced sugar in addition to cotton, I have converted sugar into cotton according to the following formula: assuming an average of 8 bales of cotton and 10 hogsheads of sugar to the hand, I multiplied the number of hogsheads of sugar by .8. The estimates come from Fort Bend County planter Jonathan Watters, "Agricultural Practice in Texas," *De Bow's Review* 26 (March 1859): 311, 313.

TABLE 10. Selected election results for Brazos counties, 1848–1859

Election	Winner	Loser
1848 presidential	Cass (D) 75%	Taylor (W) 25%
1849 gubernatorial	Bell (D) 79%	Wood (D) 21%
1851 gubernatorial	Bell (D) 77%	Epperson (W) 12%
1852 presidential	Pierce (D) 73%	Scott (W) 27%
1853 gubernatorial	Pease (D) 62%	Ochiltree (W) 28%
1855 gubernatorial	Pease (D) 52%	Dickson (A) 48%
1855 debt referendum	No 65%	Yes 35%
1856 presidential	Buchanan (D) 62%	Fillmore (A) 38%
1857 gubernatorial	Runnels (D) 58%	Houston (O) 42%
1859 gubernatorial	Houston (O) 50%	Runnels (D) 49%
1860 presidential	Breckinridge (D) 80%	Opposition 20%
1861 secession	Yes 94%	No 6%

SOURCE: Election Returns, 1851, 1852, 1853, 1855, 1856, 1857, 1859, TSL.

Appendix 2

METHOD USED TO IDENTIFY NEIGHBORHOODS

Because the emphasis here has been on neighborhood and locality, it was important that census information be gathered at the subcounty level. Unfortunately, the censuses of 1850 and 1860 often surveyed entire counties without breaking them into smaller units. To remedy this problem, I devised a method based on the assumption that those dwellings visited by the census-taker on the same day (each census form was dated) probably were near one another. Using local histories of the three neighborhoods examined (Gulf Prairie, Chappell Hill, and Cat Spring), along with church membership records and agricultural society records, I compiled a list of residents, then located them in the population census. In each instance, residents were clustered on a relatively small number of pages, approximating the geographical boundaries of the neighborhood. After locating the neighborhood cluster in the census, I entered all residents visited on the same day (or days), whether or not they had appeared in the church and agricultural society membership lists, on the assumption that they lived within short walking or riding distance from the others. In some cases, members were scattered outside the main cluster. I omitted these "outliers," assuming that they resided outside the neighborhood but, for whatever reason, retained church or organization membership. In general, a core group of longtime residents emerged over the two censuses, which helped me to check the accuracy of the method.

Breaking the counties down into localities offers many advantages over analyzing entire counties. Texas counties were often large, sometimes more than thirty miles across, and residents did not necessarily interact on a regular basis. Breaking the counties down also allows the contours of neighborhood to emerge much more clearly. Austin County, for example, was roughly divided between African Americans, Anglos, and Germans, but the population was not evenly distributed. The subcounty perspective highlights the divisions between the bottomland and planter neighborhoods, where most African Americans and Anglos lived, and the German small-farming areas in the western uplands. However, the method was not perfectly precise. Because the census-takers did not confine themselves to the boundaries of single communities, it is inevitable that some nonmembers were included and some members excluded from certain neighborhoods. In addition, even where the census identified a tract by name, the size and definition of the tract might vary from census year to census year. In some instances, I was able to correct inaccuracies through the use of tax rolls, which usually gave the rough location of an individual's land holdings. The agricultural schedules of the census occasionally enumerated farms missed in the population census, which allowed me to make further adjustments.

The sources used to locate neighborhoods and the dwelling ranges for each census are as follows (all census data is from microfilm):

Gulf Prairie

1850: U.S. Census Bureau, Free Schedule, Brazoria County, record M432, roll 908. dwellings 63–70, 246–261. I used the Brazoria County plantation map at the Brazoria County Historical Museum, which was based on Platter, "Education, Social, and Economic Characteristics of the Plantation Culture of Brazoria County."

1860: U.S. Census Bureau, Free Schedule, Brazoria County, record M653, roll 1289, dwellings 1–32; and Brazoria County plantation map, BCHM.

Chappell Hill

1850: U.S. Census Bureau, Free Schedule, Washington County, record M432, roll 916, dwellings 314–350, 405–407, 450–453, 508–510, 563–573. I used an 1853 list of the members of the Chappell Hill Methodist Church in Mr. and Mrs. Nath Winfield, *All Our Yesterdays: A Brief History of Chappell Hill* (Waco: Texian Press, 1969), 36, to identify families. I also used land-ownership data on tax rolls to eliminate some who lived closer to Washington than to Chappell Hill.

1860: U.S. Census Bureau, Free Schedule, Washington County, record M653, roll 1307, dwellings 325–390, 1046–1084, 1162–1217. Chappell Hill was listed as a census tract.

Cat Spring

1850: U.S. Census Bureau, Free Schedule, Austin County, record M432, roll

908, dwellings 250–293, 319–325. I used the Agricultural Society of Cat Spring membership roster and biographical information in the society's *The Cat Spring Story.* I also cross-checked with tax rolls.

1860: U.S. Census Bureau, Free Schedule, Austin County, record M653, roll 1287, dwellings 793–806, 1101–1201. Cat Spring was surveyed in the same tract as Industry and Welcome, so I used the Agricultural Society of Cat Spring membership roster and biographical information in *The Cat Spring Story* to locate residents of Cat Spring and Millheim.

Notes

INTRODUCTION

1. Diary Entry for December 27, 1848, in *Diary and Letters of Rutherford B. Hayes, the Nineteenth President of the United States*, vol. 1, *1834–1860*, ed. Charles Richard Williams (Columbus: Ohio State Archaeological and Historical Society, 1922), 246; Hayes to "Fanny," January 1, 1849, ibid., 247; Hayes to "Mother," January 27, 1849, ibid., 255.

2. The percentage of Africans in the slave population as a whole comes from Michael A. Gomez, *Exchanging Our Country Marks: The Transformation of African Identities in the Colonial Era and Antebellum South* (Chapel Hill: University of North Carolina Press, 1998), 23, and is adapted from Robert William Fogel and Stanley L. Engermann, *Time on the Cross: The Economics of American Negro Slavery* (New York: Norton, 1989), 23. I have estimated the African-born population of Brazoria County at between 500 and 1,000, out of a total slave population of about 3,000. For a discussion of my estimate, see chapter 1. For Brazoria County's slave population, see Randolph B. Campbell, *An Empire for Slavery: The Peculiar Institution in Texas, 1821–1865* (Baton Rouge: Louisiana State University Press, 1989), 264.

3. James F. Perry Daybook, James F. and Stephen S. Perry Papers, DBCAH. On the German population of Brazoria County, see Walter Struve, *Germans and Texans: Commerce, Migration, and Culture in the Days of the Lone Star Republic* (Austin: University of Texas Press, 1996), 72.

4. Ira Berlin, "Time, Space, and the Evolution of Afro-American Society on British Mainland North America," *American Historical Review* 85 (February 1980): 44–78.

5. Debate on the relationship between the antislavery movement and capitalism begins with Eric Williams, *Capitalism and Slavery* (Chapel Hill: University of North Carolina Press, 1944). Other landmark works include David Brion Davis, *The Problem of Slavery in the Age of Revolution, 1770–1823* (Ithaca, NY: Cornell University Press, 1975); Seymour Drescher, *Econocide: British Slavery in the Era of Abolition* (Pittsburgh: University of Pittsburgh Press, 1977); and David Eltis, *Economic Growth and the Ending of the Transatlantic Slave Trade* (New York: Oxford University Press, 1987). On Mexican abolitionism, see Robin Blackburn, *The Overthrow of Colonial Slavery, 1776–1848* (London: Verso, 1988), 368–372.

6. Jeremy Adelman and Stephen Aron, "From Borderlands to Borders: Empires, Nation-States, and the Peoples In Between in North American History," *American Historical Review* 104 (June 1999): 815–816.

7. On the connection between borders and identity, see Andrés Reséndez, *Changing National Identities at the Frontier: Texas and New Mexico, 1800–1850* (Cambridge: Cambridge University Press, 2005).

8. Adelman and Aron, "From Borderlands to Borders," 815–816.

9. Campbell, *An Empire for Slavery*, 1–2.

10. For a thorough and insightful discussion of the concept of a frontier, see Kerwin Lee Klein, *Frontiers of Historical Imagination: Narrating the European Conquest of Native America, 1890–1990* (Berkeley: University of California Press, 1997). The quotes are in David J. Weber, *The Spanish Frontier in North America* (New Haven, CT: Yale University Press, 1992), 11; William Cronon, George Miles, and Jay Gitlin, eds., *Under an Open Sky: Rethinking America's Western Past* (New York: Norton, 1992), 7. Adelman and Aron, "From Borderlands to Borders," 814–841, see frontiers as meeting places of different cultures, borderlands as the contested areas between imperial powers, and bordered lands as the areas adjacent to the boundaries of nation-states.

11. A number of otherwise excellent works take the definition of frontier for granted, and many of these ascribe Turnerian characteristics to their subjects. Eugene D. Genovese, *Roll, Jordan, Roll: The World the Slaves Made* (New York: Vintage Books, 1972), argues that Texas masters "probably deserved" their reputation for harsh treatment, but that this changed as the "frontier receded" (53); Peter Wood, *Black Majority: Negroes in Colonial South Carolina from 1670 to the Stono Rebellion* (New York: Alfred A. Knopf, 1974), sees a "pioneer existence" (p. 95) and the "egalitarian intimacies inevitable on a frontier" as having "exerted leveling influences" (96) between masters and slaves; Ira Berlin, *Many Thousands Gone: The First Two Centuries of Slavery in North America* (Cambridge, MA: Harvard-Belknap, 1998), refers to "egalitarian confrontations" between masters and slaves (66) and "sawbuck equality," amid "unsettled conditions" (67), "rude frontier conditions" and "frontier slaveowners" (68); in *Generations of Captivity: A History of African-American Slaves* (Cambridge, MA: Harvard-Belknap, 2003), 174, Berlin mentions a "pioneer plantation" (174) and "the harsh conditions of frontier life" (175); Steven F. Miller, "Plantation Labor Organization and Slave Life on the Cotton Frontier: The Alabama Black Belt, 1815–1840," in *Cultivation and Culture: Labor and the Shaping of Slave Life in the Americas*, ed. Ira Berlin and Philip Morgan (Charlottesville: University of Virginia Press, 1992), emphasizes a labor regime based on land-clearing as a defining feature of frontier life; William Dean Carrigan, "Slavery on the Frontier: The Peculiar Institution in Central Texas," *Slavery and Abolition* 20 (August 1999): 63–86, offers no definition but emphasizes sparse settlement and the presence of Indians and Mexicans as the most prominent themes. Not all historians who use the term have neglected to define it. See, for example, Daniel H. Usner Jr., *Indians, Settlers, and Slaves in a Frontier Exchange Economy: The Lower Mississippi Valley Before 1783* (Chapel Hill, NC: Omohundro Institute of Early American History and Culture, 1992), 6, who rejects Turner's "interracial boundary" definition in favor of "a network of cross-cultural interaction." See also Edward E. Baptist, *Creating an Old South: Middle Florida's Plantation Frontier before the Civil War* (Chapel Hill: University of North Carolina Press), 7, 298n.

12. The literature on the borderlands is vast. Major works include Herbert Eugene Bolton, *The Spanish Borderlands: A Chronicle of Old Florida and the Southwest* (New Haven, CT: Yale University Press, 1921); Gloria Anzaldúa, *Borderlands/Fronteras: The New Mestiza*, 2nd ed., with an introduction by Sonia Saldívar-Hull (San Francisco: Aunt Lute Books, 1999); Weber, *Spanish Frontier*. For an overview, see David J. Weber, "The Spanish Borderlands of North America: A Historiography," *Organization of American Historians Magazine of History* 14 (Summer 2000): 5–11.

13. On efforts to recapture slaves, see Campbell, *An Empire for Slavery,* 63.

14. The state capital of Austin, it should be noted, is not located in Austin County but in Travis County, approximately one hundred miles to the northwest.

PROLOGUE

1. Nan Thompson Ledbetter, "The Muddy Brazos in Early Texas," *SHQ* 63 (October 1959): 238–240.

2. The best contemporary descriptions are in Stephen F. Austin, "Journal of Stephen F. Austin on His First Trip to Texas, 1821," *Texas Historical Association Quarterly* 4 (April 1904): 286–307; *A Visit to Texas in 1831, Being the Journal of a Traveller through Those Parts Most Interesting to American Settlers* (New York: Goodrich and Wiley, 1834; Austin: Steck, 1952).

3. Robert A. Rickliss, *The Karankawa Indians of Texas: An Ecological Study of Cultural Tradition and Change* (Austin: University of Texas Press, 1996), 71–72; 99–100. See also David La Vere, *The Texas Indians* (College Station: Texas A&M University Press, 2004), 59–64.

4. Weber, *Spanish Frontier,* 147–152, 187; David La Vere, "Between Kinship and Capitalism: French and Spanish Rivalry in the Louisiana-Texas Indian Trade," *Journal of Southern History* 64 (May 1998): 201–202.

5. Elizabeth H. John, *Storms Brewed in other Men's Worlds: The Confrontation of Indians, Spanish, and French in the Southwest, 1540–1795,* 2nd ed. (Norman: University of Oklahoma Press, 1996), 551, 614, 640–641; W. W. Newcomb Jr., *The Indians of Texas: From Prehistoric to Modern Times* (Austin: University of Texas Press, 1961), 67–68; "exterminate" in Rickliss, *Karankawa Indians,* 151–152.

6. Rickliss, *Karankawa Indians,* 148.

7. Ibid., 149–150.

8. Ibid., 156; Weber, *Spanish Frontier,* 113.

9. Rickliss, *Karankawa Indians,* 171.

10. Ibid., 143–145, 159–168.

11. Ibid., 130–133.

12. Newcomb, *Indians of Texas,* 136.

13. Rickliss, *Karankawa Indians,* 137–142.

14. Donald E. Chipman, *Spanish Texas, 1519–1821* (Austin: University of Texas Press, 1992), 224–225.

15. David E. Narrett, "José Bernardo Gutiérrez de Lara: *Caudillo* of the Mexican Republic in Texas," *SHQ* 106 (October 2002): 195–228.

16. Gregg Cantrell, *Stephen F. Austin: Empresario of Texas* (New Haven, CT: Yale University Press, 1999), 75–94; 112–127.

17. Colonisation Law of 1823, in *LT,* 1:27–30.

CHAPTER 1

1. Joseph G. Baldwin, *The Flush Times of Alabama and Mississippi: A Series of Sketches,* with an introduction by James H. Justus (Baton Rouge: Louisiana State University Press, 1987), 93.

2. Jodella D. Kite, "A Social History of the Anglo-American Colonies in Mexican Texas, 1821–1835" (Ph.D. diss., Texas Tech University, 1990), 44; U.S. Census Bureau, *Population of the United States in 1860; Compiled from the Original Returns of the Eighth Census* (Washington, DC: Government Printing Office, 1864), 616–623.

3. Barnes F. Lathrop, "Migration into East Texas, 1835–1860," *SHQ* 52 (July 1948): 190; U.S. Census Bureau, Free Schedules, Washington County, Texas, 1850, record M432, roll 916; 1860, record M653, roll 1307 (microfilm).

4. D. W. Meinig, *Imperial Texas: An Interpretive Essay in Cultural Geography* (Austin: University of Texas Press, 1969), 42–46; "A Thrilling Narrative of the Adventures of a North Carolina Family While Emigrating in 1843," Theodore Davidson Morrison Papers, SHC.

5. Thomas Blackshear Diary, April 15, 1858; January 8, 1859, Thomas Edward Blackshear Papers, DBCAH; Hamblin Bass to Robert Adams, January 4, 1860, in *The Hicks-Adams-Bass-Floyd-Patillo and Collateral Lines, Together with Family Letters, 1840–1860*, ed. Gary Doyle Woods (Salado, TX: Anson Jones Press, 1963), 206–207.

6. David Eltis, "Labour and Coercion in the English Atlantic World from the Seventeenth to the Twentieth Century," *Slavery and Abolition* 14 (April 1993): 213.

7. Eltis, *Economic Growth*, 34–46.

8. For distances up- and downriver, see D. E. E. Braman, *Braman's Information about Texas* (Philadelphia: Lippincott, 1858), 164.

9. This discussion is based on maps prepared by the GLO. See maps of Brazoria and Fort Bend counties. Patricia Nelson Limerick, *The Legacy of Conquest: The Unbroken Past of the American West* (New York: Norton, 1987), 66.

10. Cantrell, *Stephen F. Austin*, 136–138; Kelly F. Himmel, *Conquest of the Karankawas and the Tonkawas, 1821–1859* (College Station: Texas A&M University Press, 1999), 50.

11. *AP*, vol. 1, pt. 2, p. 1473; quote in *A Visit to Texas in 1831*, 39.

12. Guide to Spanish and Mexican Land Grants in Texas, GLO.

13. Ibid.; J. Thomas to Stephen F. Austin, April 15, 1825, *AP*, vol. 1, pt. 1, p. 765; Contract for Hire of Slaves, ibid., 869–870; Thomas Westall to Stephen F. Austin, ibid., 941.

14. Thomas Westall Estate Inventory, Personal Correspondence, Perry Papers, DBCAH; U.S. Census Bureau, Slave Schedule, Brazoria County, Texas, 1850, record M432, roll 917 (microfilm).

15. Guide to Spanish and Mexican Land Grants, GLO; "Texas," 4:135, R. G. Dun and Company Collection, Baker Library Historical Collections, Harvard Business School; Campbell, *An Empire for Slavery*, 194; William Ransom Hogan, *The Texas Republic: A Social and Economic History* (Austin: University of Texas Press, 1969), 103.

16. Marie Beth Jones, *Peach Point Plantation: The First 150 Years* (Waco, TX: Texian Press, 1982).

17. U.S. Census Bureau, Free Schedule, Brazoria County, 1860, record M653, roll 1289; Slave Schedule, ibid., 1860, record M653, roll 1309 (microfilm).

18. For examples, see Emily Perry to James F. Perry, June 16, 1837; H. Chriesman to James F. Perry, December 7, 1838; K. K. Koontze to James F. Perry, June 26, 1839; Guy M. Bryan to James F. Perry, September 15, 1851, all in Perry Papers, DBCAH; Moses Austin Bryan to Guy M. Bryan, March 1, 1839, Guy M. Bryan Papers, DBCAH.

19. Nestor Clay to Father, April 28, 1832, Nestor and Tacitus Clay Papers, DBCAH.

20. William W. Hunter to James F. Perry, February 18, 1833, Perry Papers, DBCAH.

21. Gideon Lincecum to Brother, April 2, 1860, Gideon Lincecum Letter Press, Gideon Lincecum Papers. DBCAH.

22. Quote in J. H. Kuykendall, "Reminiscences of Early Texans: Capt. Gibson Kuykendall," *TSHAQ* 7 (July 1903): 30; Himmel, *Conquest of the Karankawas and Tonkawas*, 44–45.

23. Washington County Deeds, Book D, 344–345; Book E, 202, both in WCC. John Washington Lockhart's memoir gives 1839 or 1840 as the date of arrival for the Jacksons. See Mrs. Johnnie Wallis and L. L. Hill, eds., *Sixty Years on the Brazos: The Life and Letters of Dr. John Washington Lockhart, 1824–1900* (New York: Argonaut Press, 1966), 195. Jackson's sixteen slaves actually belonged to his wife, Julia. See Texas County Tax Rolls, Washington County, series 1, roll 637, 1837–1878, TSL; U.S. Census Bureau, Agricultural Schedule, Washington County, 1850, record T1134, roll 3; Slave Schedule, ibid., 1850, record M432, roll 918 (microfilm).

24. John Washington Lockhart, Written Remembrances, 1–2, RL.

25. Ibid.; Mr. and Mrs. Nath Winfield, *All Our Yesterdays: A Brief History of Chappell Hill* (Waco, TX.: Texian Press, 1969), 36; O. H. P. Garrett Obituary, ca. 1886, Texas Narratives, Herbert Howe Bancroft Collection, Bancroft Library, University of California, Berkeley; Washington County, Texas County Tax Rolls, series 1, roll 637, TSL. Garrett owned fifteen slaves in 1860 but paid taxes on fourteen. See U.S. Census Bureau, Slave Schedule, Washington County, 1860, record M653, roll 1312 (microfilm).

26. Winfield, *All Our Yesterdays*, 1.

27. U.S. Census Bureau, Free Schedules, Washington County, 1850, record M432, roll 916; 1860, M653, roll 1307; Slave Schedules, ibid., 1850, record 432, roll 918; 1860, record 653, roll 1312; Agricultural Schedules, ibid., 1850, record T1134, roll 3; 1860, record T1134, roll 7 (microfilm). For the method used to identify Chappell Hill residents, see appendix 2. U.S. Census Bureau, *Population of the United States in 1860*, 486, gives 440 as the free population and 1,720 as the slave population, for an 80 percent slave majority. The discrepancy arises out of different ways of defining the Chappell Hill neighborhood.

28. Wallis, *Sixty Years on the Brazos*, 17, 199; Autobiographical Sketch of Joseph Edmund Wallis, Chappell Hill Historical Society, Chappell Hill, Texas; U.S. Census Bureau, Agricultural Schedule, Washington County, 1860, record T1134, roll 7; Slave Schedule, ibid., 1860, record 653, roll 1312 (microfilm). See also John Washington Lockhart, Written Remembrances, 8–9, John Washington Lockhart Papers, RL; Account with Atkinson and Carter for 1860, Lockhart Papers, RL.

29. Michael Tadman, *Speculators and Slaves: Masters, Traders, and Slaves in the Old South* (Madison: University of Wisconsin Press, 1989), 154–155, 228.

30. Ibid., 11–12.

31. Sarah Ashley, in *The American Slave: A Composite Autobiography*, ed. George P. Rawick (Westport, CT: Greenwood Press, 1972), vol. 16, pt. 1, p. 34. Ashley's experience was unusual. She was born in Mississippi and was then purchased by a slave trader who took her to Georgia. More often, it was the reverse: Georgia slaves wound up in Mississippi. It is possible that in Ashley's memory a "Georgia trader," an older term for *slave dealer*, became transformed into a sale to Georgia.

32. Guy M. Bryan to Rutherford B. Hayes, November 18, 1847, in "The Bryan-Hayes Correspondence," ed. E. W. Winkler, *SHQ* 25 (January 1922): 199.

33. Betty Simmons, in Rawick, *The American Slave*, vol. 16, pt. 4, p. 19. No accounts of this sort exist in the WPA narratives for Brazos slaves, although that should not be taken as evidence that the interregional trade was insignificant. See the discussion that follows.

34. U.S. Census Bureau, Slave Schedules, Washington County, 1850, record 432, roll 918; 1860, record 653, roll 1312; Brazoria County, 1850, record M432, roll 917; 1860, record M653, roll 1309 (microfilm).

35. Campbell, *An Empire for Slavery*, 51. Robert William Fogel, *Without Consent or Contract: The Rise and Fall of American Slavery* (New York: Norton, 1989), 67, notes that estimates range from 16 percent to 60 percent for slaves in the Old Southwest who were sold there on the interregional market. See also Robert W. Fogel, Ralph A Galantine, and Richard L. Manning, *Without Consent or Contract: The Rise and Fall of American Slavery, Evidence and Methods* (New York: Norton, 1989), 195.

36. Tadman, *Speculators and Slaves*, 25–31, 237–247.

37. U.S. Census Bureau, *The Seventh Census of the United States: 1850* (Washington, DC: Robert Armstrong, 1853), 500–506; U.S. Census Bureau, *Population of the United States in 1860*, 478–483.

38. On the prevalence of osnaburgs in antebellum Texas, see Campbell, *An Empire for Slavery*, 139.

39. Walter D. Kamphoefner, *The Westfalians: From Germany to Missouri* (Princeton, NJ: Princeton University Press, 1987), 3–11, 70–105; Terry G. Jordan, *German Seed in Texas Soil: Immigrant Farmers in Nineteenth-Century Texas* (Austin: University of Texas Press, 1966), 30–54.

40. Peter Kreidte, Hans Medick, and Jürgen Schlumbohm, *Industrialization before Industrialization: Rural Industry and the Genesis of Capitalism*, trans. Beate Schempp (Cambridge: Cambridge University Press, 1981), 101–107, 74–82; Kamphoefner, *The Westfalians*, 17–18; Timothy G. Anderson, "Proto-Industrialization, Sharecropping, and Outmigration in Nineteenth-Century Rural Westphalia," *Journal of Peasant Studies* 29 (October 2001): 1–30.

41. Jürgen Schlumbohm, *Lebensläufe, Familien, Höfe: Die Bauern und Heuerleute des Osnabrükischen Kirschspiels Belm in proto-industrieller Zeit, 1650–1860* (Göttingen: Vandenhoeck und Ruprecht, 1994), 54–55; David Warren Sabean, *Property, Production, and Family in Neckarhausen, 1700–1870* (Cambridge: Cambridge University Press, 1990), 49, 156.

42. Friedrich Wilhelm Meier to his sister, May 12, 1866, in "*So Besint euch doch nicht lange und kommt herrüber . . .*": *Briefe von Amerikaauswanderern aus dem Kreis Lübbecke aus zwei Jahrhunderten*, ed. Heinz-Ulrich Kammeier (Darmstadt, Germany: Dissertations Druck, 1989), 36–37.

43. Wolfgang Köllmann and Peter Marschalck, "German Emigration to the United States," *Perspectives in American History* 7 (1973): 518–519.

44. Jordan, *German Seed in Texas Soil*, 50; U.S. Census Bureau, *Population of the United States in 1860*, 487–489; Struve, *Germans and Texans*, 48–49.

45. *Itzehoer Wochenblatt*, May 13, 1847; Virginia Yans-McLaughlin, ed., *Immigration Reconsidered: History, Sociology, and Politics* (New York: Oxford University Press, 1990), 7.

46. Stephanie McCurry, *Masters of Small Worlds: Yeoman Households, Gender Relations, and the Political Culture of the Antebellum South Carolina Low Country* (New York: Oxford University Press, 1995), 22–30.

47. Guide to Spanish and Mexican Land Grants, GLO; Deed, Charles W. Fordtran Papers, DBCAH; Ann Lindemann, James Lindemann, and William Richter, *Historical Accounts of Industry, Texas, 1831–1986* (New Ulm, TX: New Ulm Enterprise, 1986), 4–6.

48. Detlef Dunt, *Reise Nach Texas nebst Nachrichten von diesem Lande: Für Deutsche, welche nach Amerika zu gehen beabsichtigen* (Bremen, Germany: Carl Wilh. Wiehe, 1834), 6–12.

49. Ibid., 96; *Itzehoer Wochenblatt,* April 4, 1847; Wilhelm Richter to Parents, ca. 1847, in Heinz-Ulrich Kammeier, *"Halleluja, jetzt sehen wir Amerika": Auswandererbriefe aus dem Kreis Lübbecke und Umgebung, 1836–1889* (Espelkamp: Verlag Marie L. Leidorf, 1994), 47.

50. Jordan, *German Seed in Texas Soil,* 94–96, 157–167.

51. "Sketch of the Lives of Mary Loewenberger Amsler and Charles Conrad Amsler," typescript copy, Amsler Papers, DBCAH.

52. Ibid.; Rosa Kleberg, "Some of My Early Experiences in Texas," *TSHAQ* 1 (April 1898): 297, "pioneer life" on 299; Caroline von Hinueber, "Life of German Pioneers in Early Texas," *TSHAQ* 2 (January 1899): 228; William Trenckmann, *Austin County: Beilage zum Bellville Wochenblatt* (Bellville, TX: Bellville Wochenblatt, ca. 1899), 29–31, "bag of doubloons" on 29. The title was apparently registered in Louis von Roeder's name alone, with the Amslers receiving their lands from him.

53. Trenckmann, *Austin County,* 34–37, quote on 36; Autobiography of William Trenckmann, 3, Trenckmann Papers, DBCAH.

54. Charles Nagel, *A Boy's Civil War Story* (St. Louis: Dorrance, 1937), 39, 57.

55. "Josef L. Lesikar," in *Czech Voices: Stories from Texas in the Amerikan Narodni Kalendar,* ed. and trans. Clinton Machann and James W. Mendl Jr. (College Station: Texas A&M University Press, 1991), 13–16.

56. Friedrich Koy to Mathilde Koy, January 1, 1853; March 10, 1853, German Immigrant Letters, LOC.

57. Friedrich Koy to Mathilde Koy, March 10, 1853, German Immigrant Letters, LOC; Texas County Tax Rolls, Austin County, first series, pt. 1, reel 19, 1837–1881, TSL.

58. Jordan, *German Seed in Texas Soil,* 101–106, 115; U.S. Census Bureau, Free Schedule, Austin County, Texas, 1860, record 653, roll 1287; Agricultural Schedule, ibid., 1860, record T1134, roll 3 (microfilm).

59. Gilbert Jordan, "William Steinert's View of Texas in 1849," *SHQ* 80 (January 1977): 292–293.

60. Ammon Underwood Journal and Scrapbook, ca. 1913, BCHM. A photograph of Thompson and his wife was published in James A. Creighton, *Narrative History of Brazoria County* (Waco: Texian Press, 1975), 172.

61. There is one probable exception to this generalization. Planter Jared Groce, who arrived in present-day Austin County in 1822 with more than ninety slaves, was apparently involved in an illegal trade in 1817 while living in present-day Alabama. See U.S. Congress, Senate, *Slaves Imported by an Indian Agent Contrary to Law,* 17th Cong., 1st sess., 1822, 960–975. The number of slaves is unknown.

62. Unlike other Texas traders, who desired African labor for their own plantations, Edwards's participation seems to have been part of a larger criminal enterprise. See Sean Kelley, "Blackbirders and *Bozales:* African-Born Slaves on the Lower Brazos River of Texas in the Nineteenth Century," *Civil War History* 54 (December 2008): 410–412.

63. Eugene C. Barker, "The African Slave Trade in Texas," *Quarterly of the Texas State Historical Association* 6 (October 1902): 156–157. Fred Robbins, "The Origins and Development of the

African Slave Trade into Texas, 1816–1860" (master's thesis, University of Houston, 1972), 108, considers 500 "a conservative estimate." For Houston's speech, see Barker, "African Slave Trade," 155. See also Havana Consul to Secretary of State John Forsythe, February 12, 1836, Despatches from U.S. Consuls in Havana, February 12, 1836, to July 17, 1838, copy no. T-20, roll 7, National Archives Microfilm, RG 59; William Kennedy to Lord Aberdeen, September 5, 1843, in Ephraim Douglas Adams, "British Correspondence concerning Texas," *SHQ* 17 (October 1913): 197–201; Philip Duggan to J. T. Kirkman, August 6, 1867, Letters Sent, Subassistant Commissioner, Columbia, Texas, item 3689, RG 105, Records of the Bureau of Refugees, Freedmen, and Abandoned Lands, National Archives.

64. Price and shipping estimates for Africa and Cuba are in 1821–25 U.S. dollars. See Eltis, *Economic Growth*, 263, 280. Price estimates for Texas are in Campbell, *An Empire for Slavery*, 71. According to the Economic History Services calculator, inflation had little effect, with US$568 in 1837 worth US$549 in 1825. See http://eh.net/hmit/ppowerusd, accessed November 12, 2006.

65. Neither the original manuscript of Gray's diary nor the two published versions include the "vocabulary." See William Fairfax Gray Diaries, DBCAH; William Fairfax Gray, *From Virginia to Texas: The Diary of William Fairfax Gray* (Houston: Fletcher Young, 1965); Paul D. Lack, ed. *The Diary of William Fairfax Gray: From Virginia to Texas, 1835–1837* (Dallas: Southern Methodist University Press, 1997).

66. Eltis, *Economic Growth*, 83–86; Herbert S. Klein, "The Structure of the Atlantic Slave Trade in the Nineteenth Century: An Assessment," *Outre-Mers: Revue d'Histoire* 89 (2002): 63–77; Franklin W. Knight, *Slave Society in Cuba during the Nineteenth Century* (Madison: University of Wisconsin Press, 1970), 29, 51–54, 102–103.

67. Eltis, *Economic Growth*, 245.

68. David Eltis, Stephen D. Behrendt, David Richardson, and Herbert S. Klein, eds., *The Trans-Atlantic Slave Trade: A Database on CD-ROM* (Cambridge: Cambridge University Press, 1999).

69. William Pitt Zuber to Eugene C. Barker, March 12, 1902, Eugene C. Barker Papers, DBCAH. Zuber also alleged that the Africans "volunteered" upon being told that they would be freed after serving for three years in Texas and signed indentures to that effect. Later, when intercepted by a U.S. Navy ship, the traders presented the indentures and were allowed to continue. While this story is very unlikely, it may be a garbled account of a purchase of emancipados. The most reliable (though still questionable) biography of Monroe Edwards actually suggests that he purchased emancipados. See *Life of the Celebrated Munroe Edwards* (Boston: William White and H. P. Lewis, 1842).

70. Robin Law, *The Oyo Empire, c. 1600–1836: A West African Imperialism in the Era of the Atlantic Slave Trade* (Oxford: Clarendon Press, 1977), 307–308.

71. David Randon, *Plaintiff in Error v. Thomas Toby*, in *Cases Argued and Decided in the Supreme Court of the United States, 1850–1851*, ed. Stephen K. Williams (Rochester, NY: Lawyers Co-operative, 1901), 784–797. The "expert witness" was Thomas League, a Houston merchant who claimed to have made two trips to the Grain Coast, Gold Coast, and Slave Coast. League's expertise was discounted by the court, but the term *Lucame*, because of its widespread use in the New World, probably arose not from League's alleged travels, but in interviews either with the slaves themselves or people close to them. The court ruled that the question of whether the

Africans had been legitimately enslaved was irrelevant because they had been purchased and had lived as de facto slaves since arriving in Texas.

72. Eltis, *Economic Growth*, 169. On the prominence of Yoruba-speakers during this period, see David Eltis, "The Diaspora of Yoruba Speakers, 1650–1850," in *The Yoruba Diaspora in the Atlantic World*, ed. Toyin Falola and Matt D. Childs (Bloomington: Indiana University Press, 2004), 28–31; Paul Lovejoy, "The Yoruba Factor in the Trans-Atlantic Slave Trade," ibid, 40–51.

73. Robin Law, *Ouidah: The Social History of a West African Slaving "Port," 1727–1892* (Athens: Ohio University Press, 2004), 155–188; J. F. Ade Ajayi, "Narrative of Samuel Ajayi Crowther," in *Africa Remembered: Narratives by West Africans in the Era of the Slave Trade* ed. Philip D. Curtin (Madison: University of Wisconsin Press, 1967), 310–311; and Philip D. Curtin, "Narrative of Joseph Wright," ibid., 330.

74. Law, *Ouidah*, 18; Robin Law and Paul Lovejoy, eds., *The Biography of Mahommah Gardo Baquaqua: His Passage from Slavery to Freedom in Africa and America* (Princeton, NJ: Markus Wiener, 2001), 39.

75. Eltis et al., *The Trans-Atlantic Slave Trade*. Search terms were "25-year period": 1825–1850; "where slaves embarked": Bight of Benin; "region in which greatest number of slaves disembarked": Cuba. On mortality see Eltis, *Economic Growth*, 137.

76. Eltis, *Economic Growth*, 165, 169; Klein, "Structure of the Atlantic Slave Trade," 65; Joseph C. Miller, *Way of Death: Merchant Capitalism and the Angolan Slave Trade, 1730–1830* (Madison: University of Wisconsin Press, 1988), 507–508, 528–529, 649–650; Affidavits of Singweh [Cinque] and Kimbo, October 7, 1839, Lewis Tappan Papers, LOC, online versions at http://lcweb2.loc.gov/ammem/aaohtml/exhibit/aopart1b.html#0113–0128, accessed September 2, 2009.

77. Joseph C. Miller, "Central Africa during the Era of the Slave Trade, c. 1490s-1850s," in *Central Africans and Cultural Transformations in the American Diaspora*, ed. Linda M. Heywood (Cambridge: Cambridge University Press, 2002), 57–59.

78. Ibid., 22; Miller, *Way of Death*, 23–26, 225–226.

79. Eltis, *Economic Growth*, 165–168, 171–173.

80. Locations of plantations are taken from Allen A. Platter, "Educational, Social, and Economic Characteristics of the Plantation Culture of Brazoria County, Texas" (Ph.D. diss., University of Houston, 1961); Creighton, *Narrative History of Brazoria County*, 479–497; P. A. Champomier, *Statement of the Sugar Crop Made in Louisiana* (New Orleans: Cook, Young, 1853), 44. For documentation of the African presence at Peach Point, see James F. Perry, Plantation Cotton Accounts and Daybook, Perry Papers, DBCAH; for the Mills's Low Wood, see Oscar Addison Papers, Private Journal, February 21, 1861, DBCAH; for the John Greenville, Pleasant, and Leander H. McNeel plantations, see William Pitt Zuber to Eugene C. Barker, March 12, 1902, Barker Papers, DBCAH; for the Peter Bertrand Plantation, see Bertrand Estate Inventory, Probate Files, BCC; for the Fannin-Mims Plantation, see James Fannin Estate Inventory, Probate Files, BCC; for the William J. Russell Plantation, see *Edwards v. Russell*, file no. L96.01.1184, BCHM; for Charles D. Sayre, see clipping from *Galveston Daily News*, May 10, 1936, Black History File, BCHM; for Monroe Edwards's Chenango and Sterling McNeel's Darrington, see William S. Fisher to S. R. Fisher, March 2, 1836, box 2-9/21, series 73, TSL; for Warren D. C. Hall's China Grove, see *Planter's Gazette*, February 4, 1843; for William J. Hill's Osceola, see *Monroe Edwards v.*

William J. Hill, file L96.01.1208, BCHM; for Edwin Waller, see *Brazoria Brazos Courier,* December 1, 1840. Mills owned two other plantations in the county in addition to Low Wood. It is likely that African-born slaves lived on them, but there is no direct evidence of it. In addition to the preceding, there are several owners of African-born slaves who apparently lived in Brazoria County but whom I was unable to locate geographically. For Solomon Page, see *Monroe Edwards v. Solomon Page,* file L96.01.1187, BCHM; for T. F. L. Parrott and Peyton Splane, see *Monroe Edwards v. TFL Parrott,* file L96.01.1135, BCHM; for James Cleavland, see *Monroe Edwards v. James Cleavland,* file L96.01.1126, BCHM. For an account of Africans conversing with one another, see Lack, *Diary of William Fairfax Gray,* 151. Gray also states that the Africans did not speak or understand Spanish.

81. E. P. Thompson, *The Making of the English Working Class* (New York: Vintage Books, 1966), 9.

CHAPTER 2

1. Seth Ingram to Stephen F. Austin, November 27, 1824, in *AP,* vol. 2, pt. 1, p. 977. Austin apparently rejected Tone's argument, although the reasons are unclear. The 1826 census lists Ingram as living alone, which suggests that his "family," if indeed he had one, either never came to Texas or soon left. He was granted a half sitio as Jamison's partner. See Lester G. Bugbee, "The Old Three Hundred: A List of Settlers in Austin's First Colony," *Quarterly of the Texas State Historical Association* 1 (October 1897): 113, 117.

2. McCurry, *Masters of Small Worlds.*

3. Richard Boyer, *Lives of the Bigamists: Marriage, Family, and Community in Colonial Mexico,* abridged ed. (Albuquerque: University of New Mexico Press, 2000), 61–99; Steve J. Stern, *The Secret History of Gender: Women, Men, and Power in Late Colonial Mexico* (Chapel Hill: University of North Carolina Press, 1995).

4. Moses Austin's Application for Colonization Permit, December 26, 1820, in *AP,* vol. 1, part 1, p. 372; Colonization Law of 1823, in *LT,* 1:30; Colonization Law for Coahuila and Texas, in *LT,* 1:42.

5. Bugbee, "The Old Three Hundred," 108–120; Austin Colony Census of 1826, GLO.

6. Noah Smithwick, *The Evolution of a State; or, Recollections of Old Texas Days* (Austin: University of Texas Press, 1983), 16; Agreement between William Cooper and John Bird, December 29, 1832, Colonial Records, Book A, ACC; Austin Colony Census of 1826, GLO.

7. *Patton Heirs v. Columbus R. Patton,* Record no. 453, BCC.

8. Smithwick, *Evolution of a State,* 46; Stephen F. Austin to Father Juan Nepomuceno Peña, August 26, 1824, in *AP,* vol. 2, pt. 2, p. 1033.

9. "Ann Raney Thomas Coleman," in *Texas Tears and Texas Sunshine: Voices of Frontier Women,* ed. Jo Ella Exley (College Station: Texas A&M University Press, 1985), 31–35. On Texas women finding partners, see Angela Boswell, *Her Act and Deed: Women's Lives in a Rural Southern County, 1837–1873* (College Station: Texas A&M University Press, 2001), 33.

10. U.S. Census Bureau, Free Schedules, Brazoria County, Texas, 1850, record 432, roll 908; 1860, record 653, roll 1289; ibid., Austin County, Texas, 1850, record 432, roll 908; 1860, record

653, roll 1287; ibid., Washington County, Texas, 1850, record 432, roll 916; 1860, record 653, roll 1307 (microfilm).

11. *Patton Heirs v. Columbus R. Patton*, BCC.

12. Mark M. Carroll, *Homesteads Ungovernable: Families, Sex, Race, and the Law in Frontier Texas, 1823–1860* (Austin: University of Texas Press, 2001), 120–125.

13. For a detailed discussion of Texas law, see Boswell, *Her Act and Deed*, 12–30.

14. Emily Perry to James F. Perry, June 16, 1837, Perry Papers, DBCAH.

15. James F. Perry to Emily Perry, May 25, 1851, Perry Papers, DBCAH.

16. James F. Perry to Stephen F. Perry, February 29, 1844, Perry Papers, DBCAH; Thomas Affleck to Waterer and Godfrey, 468, box X-59, letter book 2, carbon copies 1–100, February–July 1854, Thomas Affleck Papers, HML.

17. Petition, Records of the Governor, Peter Hansbrough Bell, box 301-19, folder 12, TSL.

18. John Washington Lockhart to Sallie Lockhart, February 14, 1863, Lockhart Papers, RL.

19. Christine Leigh Heyrman, *Southern Cross: The Beginnings of the Bible Belt* (New York: Alfred A. Knopf, 1997), 125–129, 158–160, 166–168, 190–205; Lizzie Scott Neblett to Alice Scott, August 20, 1860, Lizzie Scott Neblett Papers, DBCAH; Rebecca Pilsbury Diary, November 14, 1848, SHC; *Anderson Texas Baptist*, October 10, 1855.

20. *Columbia Planter*, March 29, 1844.

21. Will Neblett to Lizzie Scott Neblett, January 23, 1854, Neblett Papers, DBCAH.

22. Jennie Teel to Lizzie Scott Neblett, January 2, 1860, Neblett Papers, DBCAH.

23. In her study of nearby Colorado County, Angela Boswell found that women of all classes filed for divorce. See Boswell, *Her Act and Deed*, 51. I was only able to identify one slaveholder among those filing in Washington County, and none of the petitions I found mentioned ownership of slaves. Of course, if planter and yeoman women filed for divorce at comparable rates, it would lend support to the notion that they shared similar ideas of family.

24. *Nixon v. Nixon*, DCAC.

25. Brenda E. Stevenson, *Life in Black and White: Family and Community in the Slave South* (New York: Oxford University Press, 1996), 140–141.

26 *LT*, 2:483–486. In 1848 the Texas Supreme Court narrowed the definition of cruelty to threats on the life or health of a spouse. See Boswell, *Her Act and Deed*, 49.

27. *Rachel H. Crabtree v. James B. Crabtree*, Final Record, Book B, 49, DCWC; *Livingston Wilkinson v. Mahala Wilkinson*, Final Record, Book C, 83, DCWC; *Elijah Pennington v. Sarah Pennington*, Final Record, Book F, 70, DCWC.

28. *Mary Thompson v. John Thompson*, Final Record, Book O, 49, DCWC.

29. *Mary Wells v. Edward Wells*, Final Record, Book K, 145, DCWC.

30. *Mary J. Starnes v. Arthur A. Starnes*, Final Record, Book F, 257, DCWC.

31. *Sophia Lewis v. William W. Lewis*, Final Record, Book M, 392, DCWC.

32. Robert I. Burns, S.J., ed., *Las Siete Partidas*, trans. Samuel Parsons Scott, vol. 4, *Family, Commerce, and the Sea: The Worlds of Women and Merchants* (Philadelphia: University of Pennsylvania Press, 2001), 901; Colin A. Palmer, *Slaves of the White God: Blacks in Mexico, 1570–1650* (Cambridge, MA: Harvard University Press, 1976), 85–86.

33. Campbell, *An Empire for Slavery,* 21.

34. *Monroe Edwards v. William J. Russell et al.*, February 19, 1838, case number L96.01.1184, BCHM.

35. Austin Colony Census of 1826, GLO. The manuscript census does not break the colony down by locality. I was able to determine the locations of 235 out of 290 households by cross-checking the following sources: Bugbee, "The Old Three Hundred," 108–117; *The Handbook of Texas Online* (Austin: Texas State Historical Association, 1999–2003), www.tsha.utexas.edu/handbook/online/index.new.html; County Land Grant Maps for Brazoria, Fort Bend, Austin, Grimes, and Washington Counties, GLO.

36. Jared Groce Slave Inventory, 1831, Groce Family Papers, DBCAH; R. H. and Catherine Williamson Divorce Settlement, Colonial Records, Book A, ACC; Henry William Munson Estate Records, 1833–1860, Munson Family Papers, DBCAH.

37. Constitution of the Republic of Texas, in *LT,* 1:1079; An Act to Permit Free Persons of African Descent to Select Their Own Master and Become Slaves, in *LT,* 4:947–948.

38. U.S. Census Bureau, *The Seventh Census of the United States,* 500–503; U.S. Census Bureau, *Population of the United States in 1860,* 478–483. On Loudoun County, see Stevenson, *Life in Black and White,* 208, 395.

39. Campbell, *An Empire for Slavery,* 156; Ann Patton Malone, *Sweet Chariot: Slave Family and Household Structure in Nineteenth-Century Louisiana* (Chapel Hill: University of North Carolina Press, 1992), 16–17, 271–272. Because Malone's quantifications depend so heavily on probate inventories, a similar work for Texas is probably impossible, for the simple reason that slavery was widespread for only one generation and most slave owners died only after emancipation. See also Stevenson, *Life in Black and White,* 325.

40. Peach Point Ledger, Perry Papers, DBCAH.

41. James Fannin Inventory 162, Probate Files; Joseph Mims Inventory 347, Probate Files; and Sarah Mims Inventory, Wills Book C, all in BCC; U.S. Census Bureau, Slave Schedule, Brazoria County, Texas, 1850, record M432, roll 917; Population Schedule, ibid., 1870, record M593, roll 1576 (microfilm). On the succession of the estate, see the Fannin and Mims probate records and Creighton, *Narrative History of Brazoria County,* 88.

42. James Fannin Inventory 162, Probate Files; Joseph Mims Inventory 347, Probate Files; and Sarah Mims Inventory, Wills Book C, all in BCC; U.S. Census Bureau, Population Schedule, Brazoria County, 1870, record M593, roll 1576 (microfilm). I have located no evidence in the antebellum documents of African women marrying creole men. For more instances of African-born women married to African-born men, see the sale notice in *Columbia Planter's Gazette,* February 4, 1843.

43. Probate Files 162, 347; Wills, Book C, both in BCC. I have linked those with the 1850 Slave Schedules, which did not name the Mims slaves but did record sex and age in what appeared to be groupings of mothers with their children. I used the spacing and ages of the children to match.

44. James P. Caldwell to Gerard B. Munson, June 8, 1854, Munson Family Papers, DBCAH.

45. Plantation Cotton Accounts and Day Book; Peach Point Plantation Ledger, both in Perry Papers, DBCAH; Thomas Blackshear Diary and Memorandum Book, vol. 3, Blackshear Papers, DBCAH. On independent production in general, see Ira Berlin and Philip D. Morgan, eds., *The*

Slaves' Economy: Independent Production by Slaves in the Americas (London: Frank Cass, 1991). On Louisiana, see Roderick A. McDonald, *The Economy and Material Culture of Slaves: Goods and Chattels on the Sugar Plantations of Jamaica and Louisiana* (Baton Rouge: Louisiana State University Press, 1993); Richard Follett, *The Sugar Masters: Planters and Slaves in Louisiana's Cane World, 1820–1860* (Baton Rouge: Louisiana State University Press, 2005), 123–128. For a comparison of cotton and sugar plantations, see Berlin, *Generations of Captivity*, 177–181. Finally, compare the findings for cotton plantations in Larry E. Hudson, *To Have and to Hold: Slave Work and Family Life in Antebellum South Carolina* (Athens: University of Georgia Press, 1997), especially the material in the appendix, with those of McDonald, *Economy and Material Culture of Slaves.*

46. Oyeronke Oyewumi, *The Invention of Women: Making an African Sense of Western Gender Discourses* (Minneapolis: University of Minnesota Press, 1997).

47. Kevin Roberts, "Yoruba Family, Gender, and Kinship Roles in New World Slavery," in Falola and Childs, *The Yoruba Diaspora in the Atlantic World*, 250–252.

48. Plantation Cotton Accounts and Daybook and Peach Point Plantation Ledger, Perry Papers, DBCAH.

49. Jordan, "Steinert's View of Texas in 1849," 81 (July 1977): 61.

50. Jurgen Schlumbohm, "From Peasant Society to Class Society: Some Aspects of Family and Class in a Northwest German Protoindustrial Parish, 17th–19th Centuries," *Journal of Family History* 17 (April 1992): 183–199.

51. Sabean, *Property, Production, and Family in Neckarhausen*, 122.

52. Autobiography of William A. Trenckmann, 21, William A. Trenckmann Papers, DBCAH. See also Lauren Ann Kattner, "The Diversity of Old South White Women: The Peculiar Worlds of German American Women," in *Discovering the Women in Slavery: Emancipating Perspectives on the American Past*, ed. Patricia Morton (Athens: University of Georgia Press, 1996), 299–311.

53. Autobiography of William A. Trenckmann, 22, DBCAH.

54. Nagel, *A Boy's Civil War Story*, 123.

55. Peter Borscheid, "Romantic Love or Material Interest: Choosing Partners in Nineteenth-Century Germany," *Journal of Family History* 11 (April 1986): 157–168.

56. Linda K. Kerber, *Women of the Republic: Intellect and Ideology in Revolutionary America* (Chapel Hill: University of North Carolina Press, 1980).

57. Nagel, *A Boy's Civil War Story*, 95.

58. Johann Friedrich Winkelmann [to his father], ca. 1863, reprinted in Heinz-Ulrich Kammeier, *"Ich muß mir Ärgern, das ich nicht ehr übern Großen Ozean gegangen bin": Auswanderer aus dem Kreis Lübbecke und Umgebung berichten aus Amerika* (Espelkamp: Mittwalddruck Espelkamp, 1988), 177.

59. Autobiography of William A. Trenckmann, 22, DBCAH.

60. *Anna Pohl v. Anton Pohl*, Suit 755, DCAC; *Frederica Krantzmann v. William Krantzmann*, Suit 913, DCAC; *Cath. Friedlander v. Jacob Friedlander*, Suit 914, DCAC; Mathilde Koy to Franz Koy, undated, 1853, German Immigrant Letter Çollection, LOC.

61. Deposition of Sophia Seeburger, *Barbara Meury v. Samuel Meury*, Suit 958, DCAC.

62. *Wilhelmina Buntzel v. Frederic Buntzel*, Suit 724, DCAC.

63. *Louisa Maria Luhn v. Frederick William Luhn*, Suit 437, DCAC; *Frederike Gaul v. Frederick Gaul*, Suits 395, 475, DCAC.

64. *Joseph Habermacher v. Elisabeth Habermacher,* Suit 380, DCAC.

65. *Christian Streapa v. Sophia Streapa,* Suit 737, DCAC.

66. Boswell, *Her Act and Deed,* 52.

67. David Warren Sabean, *Property, Production, and Family in Neckarhausen, 1700–1870* (New York: Cambridge University Press, 1990), 88–123.

CHAPTER 3

1. José María Sánchez, "A Trip to Texas in 1828," trans. Carlos E. Castaneda, *SHQ* 29 (April 1926): 271; Frederick Law Olmsted, *A Journey through Texas; or, A Saddle Trip on the Southwestern Frontier* (Austin: University of Texas Press, 1978), 116–117.

2. Here I borrow the model (though not the terminology) of Stuart Hall. Hall has emphasized that ideology does not "belong" to any one group and is not imposed in such a way that subaltern groups might internalize it. Rather, it is discursive and linguistic in nature and may (and he emphasizes the nondeterministic nature of this process) form articulations (or connections) with other ideas. To avoid confusion with other definitions of ideology, which tend to imply ownership by the dominant group, I have chosen to use the term *language.* For a complete discussion, see Stuart Hall, "The Problem of Ideology: Marxism without Guarantees," 40; and Lawrence Grossberg, ed., "On Postmodernism and Articulation: An Interview with Stuart Hall," 144, both in *Stuart Hall: Critical Dialogues in Cultural Studies,* ed. David Morley and Kuan-Hsing Chen (London: Routledge, 1996).

3. The starting point for all discussions of paternalism is Genovese, *Roll, Jordan, Roll,* 3–7 and passim. Note also that the term has often been used simultaneously to describe a prebourgeois mode of production in conflict with liberal capitalism. For the purposes of this discussion, paternalism refers only to the language of master-slave negotiation and not to the slave mode of production. For a dissenting view on paternalism as a whole, see James Oakes, *The Ruling Race: A History of American Slaveholders* (New York: Vintage Books, 1982); and James Oakes, *Slavery and Freedom: An Interpretation of the Old South* (New York: Vintage Books, 1990). For a recent discussion of paternalism that studiously avoids the implication that slaves internalized the paternalist ethos, see Walter Johnson, *Soul by Soul: Life inside the Antebellum Slave Market* (Cambridge. MA: Harvard University Press, 1999), 36, 206, 232.

4. Willie Lee Rose, "The Domestication of Domestic Slavery," in *Slavery and Freedom,* expanded ed., ed. William W. Freehling (New York: Oxford University Press, 1982), 18–36; Philip Morgan, *Slave Counterpoint: Black Culture in the Eighteenth-Century Chesapeake and Lowcountry* (Chapel Hill, NC: Omohundro Institute of Early American History and Culture, 1998), 284–296.

5. Eliza Perry to Stephen S. Perry, October 16, 1843, Perry Papers, DBCAH.

6. Mary Austin Holley, *Texas* (Austin: Steck, 1935), 106; Richard Slotkin, *The Fatal Environment: The Myth of the Frontier in the Age of Industrialization, 1800–1890* (New York: Atheneum, 1985; Harper Perennial, 1994).

7. Useful discussions of this issue appear in Edward W. Said, *Orientalism* (New York: Vintage Books, 1979); Joan Wallach Scott, *Gender and the Politics of History* (New York: Columbia Univer-

sity Press, 1988); Kathleen M. Brown, *Good Wives, Nasty Wenches, and Anxious Patriarchs: Gender, Race, and Power in Colonial Virginia* (Chapel Hill, NC: Institute for Early American History and Culture, 1996).

8. Drew Gilpin Faust, "Introduction: The Proslavery Argument in History," in *The Ideology of Slavery: Proslavery Thought in the Antebellum South, 1830–1860*, ed. Drew Gilpin Faust (Baton Rouge: Louisiana State University Press, 1981), 3.

9. Faust, *Ideology of Slavery*, "*Servitude*," 95; "slavery" to "roam the earth," 81–82; "savage life" to "rudest," 92; "When [the Israelites]," 61.

10. Cantrell, *Stephen F. Austin*, 189–190.

11. Stephen F. Austin to W. H. Wharton, April 24, 1829, in *AP*, 2:212.

12. Stephen F. Austin to Ramón Músquiz and Lorenzo de Zavala, July 23, 1829, in *AP*, 2:239; quoted in Campbell, *An Empire for Slavery*, 28.

13. Lizzie Atkins Narrative in Rawick, *The American Slave*, Supplement Series 2, Texas Narratives, pt. 1, p. 93. For other examples, see Ronnie C. Tyler, "Fugitive Slaves in Mexico," *Journal of Negro History* 57 (January 1972): 3.

14. William C. C. Claiborne to Simón de Herrera, August 26, 1806, Claiborne Manuscripts, HML.

15. Campbell, *An Empire for Slavery*, 16–25.

16. Sean Kelley, "'Mexico in His Head': Slavery and the Texas-Mexico Border, 1810–1860," *Journal of Social History* 37 (Spring 2004): 708–723. Interestingly, two Texas Supreme Court cases addressed, post-facto, the legal status of slavery in Mexican Texas. Neither, however, originated on the Brazos. In *Guess v. Lubbock*, 5 Tex. 535 (1851), the court upheld the claim to freedom of a woman who had been manumitted before the implementation of the Texas Constitution on March 17, 1836, commenting that Mexican antislavery laws and decrees "show conclusively a determined hostility to the institution of slavery." In *Honey v. Clark*, 37 Tex. 686 (1873), the court upheld the legitimacy of the heirs of a white slave owner who had married a slave woman before independence in 1836, finding that "the institution of slavery. . . could not exist by law until after Texas established her independence and declared herself a slave state."

17. John "Rip" Ford, a journalist, legislator, and filibuster, estimated the number at 3,000 in 1851 and 4,000 in 1855. See John Salmon Ford, *Rip Ford's Texas*, edited and with an introduction and commentary by Stephen B. Oates (Austin: University of Texas Press, 1963), 196; Tyler, "Fugitive Slaves in Mexico," 6.

18. Olmsted, *A Journey through Texas*, 324–325. For a similar statement by a slave, see Benjamin Lundy, *Life, Travels, and Opinions of Benjamin Lundy*, compiled by Thomas Earle (New York: Arno Press, 1969), 36.

19. U.S. Census Bureau, Slave Schedule, Washington County, Texas, 1860, record M653, roll 1312 (microfilm). Census statistics on fugitive slaves are notoriously inaccurate, to the point of worthlessness. The phrasing of the census question regarding fugitives was usually interpreted very narrowly to apply only to those slaves still missing and known to have left the state. Failed attempts to flee were generally not counted. Southerners, including the compiler of the 1850 census, Louisianian J. D. B. De Bow, used the low figures to demonstrate slave contentment. See John Hope Franklin and Loren Schweninger, *Runaway Slaves: Rebels on the Plantation* (New York: Oxford University Press, 1999), 279.

20. Ford, *Rip Ford's Texas*, 196. The estimate for the slave population of 1855 is based on an interpolation of figures in U.S. Census Bureau, *Statistical View of the United States* (Washington, D.C.: A. O. P Nicholson, 1854), 308–310; and U.S Census Bureau, *Population of the United States in 1860*, 483.

21. Guy M. Bryan to James F. Perry, September 15, 1851, Perry Papers, DBCAH.

22. B. J. White to Stephen F. Austin, October 17, 1835, in *AP*, 3:190.

23. *Galveston News*, September 11, 1856 (parentheses in original.) See also Neil Foley, *The White Scourge: Mexicans, Blacks, and Poor Whites in Texas Cotton Culture* (Berkeley: University of California Press, 1997), 25.

24. S. M. Swenson to John Adriance, September 1844 (no day indicated), September 16, 1844, John Adriance Papers, DBCAH. According to a Mexican statute of 1827, any slave who "for the sake of convenience, shall wish to change his master, shall be permitted to do so, provided the new master indemnify the former for what the slave cost him agreeable to the conveyance." See *LT*, 1:202. Whether Garbo's master had this obsolete statute in mind when he sent him out is impossible to say, but it does suggest another possible way in which Mexican precedent may have influenced the master-slave relationship following independence.

25. *Brazoria Brazos Courier*, December 1, 1840.

26. Berlin and Morgan, *Cultivation and Culture*, 1–2. On negotiation in general, ibid., 6; Berlin, *Generations of Captivity*, 4–5; Johnson, *Soul by Soul*, 187.

27. See, for example, Stephen F. Austin to Josiah H. Bell, January 18, 1824, in *AP*, 1:722; Holley, *Texas*, 124.

28. Smithwick, *Evolution of a State*, 4.

29. Terry G. Jordan and Matti Kaups, *The American Backwoods Frontier: An Ethnic and Ecological Interpretation* (Baltimore: Johns Hopkins University Press, 1989), 94–100.

30. Friedrich Schlecht, *On to Texas! The Story of a Journey to Texas in 1848*, English translation and notes by Charles Patrick (Manor, TX: Indio Bravo Press, 1998), 96.

31. Austin Colony Census of 1826, GLO.

32. Lundy, *Life, Travels, and Opinions*, 38; Arthur Hennis to James F. Perry, April 22, 1838, Perry Papers, DBCAH.

33. Lack., *Diary of William Fairfax Gray*, 219–220; Ann Rainey Thomas Coleman Narrative, in Exley, *Texas Tears and Texas Sunshine*, 32.

34. U.S. Census Bureau, *The Seventh Census of the United States*, 514–515; U.S. Census Bureau, *Agriculture of the United States in 1860* (Washington, DC: Government Printing Office, 1864), 140–148.

35. Address by Austin to Colonists, in *AP*, 1:819–820.

36. Hogan, *The Texas Republic*,16; Juan N. Almonte, "Statistical Report on Texas," C. E. Castañeda, trans., *SHQ* 28 (January 1925): 199. Hogan suggests that Almonte's estimate is actually too high.

37. Berlin, *Generations of Captivity*, 178–179.

38. On the "agricultural revolution" of the 1840s, see John Hebron Moore, *The Emergence of the Cotton Kingdom in the Old South: Mississippi, 1770–1860* (Baton Rouge: Louisiana State University Press, 1988), 21, 37–56. The revolution would have been less dramatic in Texas, where plantation agriculture was less developed. See Jonathan Watters, "Agricultural Practice in Texas,"

New Orleans De Bow's Review 26 (March 1859): 310–311. Plowing and planting began in Natchez around the end of March. See Ulrich B. Phillips, ed., *Plantation and Frontier, 1649–1863* (Cleveland: Arthur H. Clark, 1910), 2:234.

39. Berlin, *Generations of Captivity*, 177–179; Thomas Blackshear Diary, Blackshear Papers, DBCAH.

40. Thomas Blackshear Diary, Blackshear Papers, DBCAH; See also James F. Perry Diary, Perry Papers, DBCAH.

41. Thad Sitton and Dan K. Utley, *From Can See to Can't: Texas Cotton Farmers on the Southern Prairies* (Austin: University of Texas Press, 1997), 205–221; Solomon Northup, *Twelve Years a Slave* (New York: Dover, 1970), 178.

42. Northup, *Twelve Years a Slave*, 165–168, provides the best contemporary description of picking from the slave's point of view. From the planter's point of view, several Brazos picking records survive. See Cotton Accounts, Perry Papers, DBCAH; Cotton Picking Record Fragment, William W. Browning Papers, DBCAH; John Washington Lockhart Diary, 1859, Lockhart Papers, box 3, folder 1, RL. Lockhart's amounts represent daily totals for all of his hands, as recorded in a pocket diary. Presumably he, like the others, kept more detailed ledgers as well.

43. Richard King, ed., *Victorian Lady on the Texas Frontier: The Journal of Ann Raney Coleman* (Norman: University of Oklahoma Press, 1971), 77. On "tasking" outside of Carolina, see James O. Breeden, ed., *Advice among Masters: The Ideal of Slave Management in the Old South* (Westport, CT: Greenwood, 1980), 21, 61, 68, 73–74.

44. Northup, *Twelve Years a Slave*, 167–168.

45. Peach Point Plantation Cotton Accounts and Daybook, Perry Papers, DBCAH; *Elizabeth, fwc v. Masterson*, July 28, 1866, Letters Sent, Columbia, Texas, Item 3689, RG 105, Records of the Bureau of Refugees, Freedmen, and Abandoned Lands, National Archives.

46. Peach Point Plantation Cotton Accounts and Daybook, Perry Papers, DBCAH.

47. See entries for December 4, 1848, and December 26, 1848, Rebecca Pilsbury Diary, SHC; M. H. Washington to Governor P. H. Bell, June 27, 1852, Records of the Governor, Peter Hansborough Bell, box 301-21, folder 24, TSL.

48. William Bradbury Diary (Transcript), DBCAH, 69 and passim.

49. Smithwick, *Evolution of a State*, 16–18.

50. Confession of Charles, Black History Collection, DBCAH.

51. Brazoria County Tax Roll, 1837, Texas County Tax Rolls, Brazoria County, pt. 1, 1837–1884, microfilm edition.

52. Virtually all of the slaves imported from Cuba were African-born. The 1870 census, the first to give birthplaces for former slaves, listed 120 Africans and only 1 Cuban in Brazoria County. In addition, observers in the 1830s emphasized that the Africans did not speak Spanish. See Lack, *Diary of William Fairfax Gray*, 141.

53. On sugar plantations in general, see J. Carlyle Sitterson, *Sugar Country: The Cane Sugar Industry in the South, 1753–1950* (Lexington: University of Kentucky Press, 1953); Follett, *The Sugar Masters*.

54. Watters, "Agricultural Practice in Texas," 311; Sitterson, *Sugar Country*, 114. See also James F. Perry Plantation Diary, 1848, Perry Papers, DBCAH.

55. Watters, "Agricultural Practice in Texas," 311–312; entry for May 25, 1866, Sarah McNeill

Diary (typescript), Levi Jordan Papers, private collection. Special thanks to Ginny Raska for allowing me to examine the diary.

56. Stephen S. Perry to McLeod, August 14, 1859, Perry Papers, DBCAH.

57. Abner Jackson Estate Inventory, July 1, 1862, no. 764, BCC; Accounting of Work Done by Simon Neece, ca. 1860, Perry Papers, DBCAH.

58. Sitterson, *Sugar Country*, 17.

59. Distribution of Hands in Sugar Rolling, 1857, John H. Randolph Papers, in Kenneth Stampp, ed., *Records of the Ante-Bellum Southern Plantations from the Revolution through the Civil War*, Series 1, Selections from Louisiana State University, pt. 1, reel 14, HML. Randolph moved operations to the Brazos, to Washington County, during the Civil War but did not attempt to grow sugar.

60. Distribution of Hands, Randolph Papers, HML; Sitterson, *Sugar Country*, 140.

61. Sitterson, *Sugar Country*, 140.

62. Ibid.

63. It was difficult to divide sugar-making into tasks, but not impossible. Cane-cutting in Bahia operated on a task system. See Stuart B. Schwartz, *Sugar Plantations in the Formation of Brazilian Society: Bahia, 1550–1835* (Cambridge: Cambridge University Press, 1985), 140–141.

64. Peach Point Plantation Cotton Accounts and Daybook, Perry Papers, DBCAH.

65. Smithwick, *Evolution of a State*, 28, 49; Entry for October 24, 1839, Joseph P. Sneed Journal, Addison Papers, DBCAH.

66. Austin Colony Census of 1826, GLO. On the possibility of African slaves at Groce's, see U.S. Congress, Senate, *Slaves Imported by an Indian Agent*, 960–975.

67. "Personal Recollections of Leonard Waller Groce, as related to his son, William Wharton Groce," Groce Family Papers, DBCAH; Bill for Ferriage, 1860, Personal Correspondence, Perry Papers, DBCAH; *R. and D. G. Mills v. Ashe and Wife*, in Helen Tunnicliff Catterall, ed., *Judicial Cases concerning American Slavery and the Negro* (New York: Octagon Books, 1968), 5:292–293.

68. U.S. Census Bureau, *Population of the United States in 1860*, 503; U.S. Census Bureau, *Statistical View of the United States*, 522–525. On the planters' suspicion of itinerant preachers, see entries for February 12, 1861, and June 2, 1861, Diary of Oscar Addison, Addison Papers, DBCAH.

69. Lorena S. Walsh, *From Calabar to Carter's Grove: The History of a Virginia Slave Community* (Charlottesville: University of Virginia Press, 1997), 137.

70. Kenneth L. Brown and Doreen C. Cooper, "Structural Continuity in an African-American Slave and Tenant Community," *Historical Archaeology* 24, no. 4 (1990): 7–19; quotes in Kenneth L. Brown, "Material Culture and Community Structure: The Slave and Tenant Community at Levi Jordan's Plantation, 1848–1892," in *Working toward Freedom: Slave Society and Domestic Economy in the American South*, ed. Larry Hudson Jr. (Rochester, NY: University of Rochester Press, 1994), 114, 112. On *minkisi* (the plural of *nkisi*), see John K. Thornton, "Religious and Ceremonial Life in the Kongo and Mbundu Areas, 1500–1750," in Heywood, *Central Africans and Cultural Transformations*, 72–90.

71. Robert Farris Thompson, *Flash of the Spirit: African and Afro-American Art and Philosophy* (New York: Vintage, 1984), 18–19, 52–55. For a similar conclusion, see Walsh, *From Calabar to Carter's Grove*, 106–108, 133, 169, 202–203.

72. Patsy Moses, in Rawick, *The American Slave*, Texas Narratives, series 2, vol. 7, pt. 7, pp. 2782–2784; Lucy Lewis, ibid., Texas Narratives, vol. 16, pt. 3, p. 16. On the use of amulets by African Muslims, see João José Reis, *Slave Rebellion in Brazil: The Muslim Uprising of 1835 in Bahia* (Baltimore: Johns Hopkins University Press, 1993), 98–104.

73. Cinto Lewis, in Rawick, *The American Slave*, Texas Narratives, vol. 16, pt. 3, p. 2.

74. Brazoria County Tax Rolls, 1837; Nancy Antwine, in Rawick, *The American Slave*, Texas Narratives, series 2, vol. 2, pt. 1, p. 61.

75. Gomez, *Exchanging Our Country Marks*, 219–220.

76. U.S. Census Bureau, *Population of the United States in 1860*, 471–474; U.S. Census Bureau, *Statistical View of the United States*, 522–525; Lizzie Atkins, in Rawick, *The American Slave*, Texas Narratives, series 2, pt. 1, p. 97; Steve Robertson, ibid., supplement 2, vol. 8, p. 3335.

77. Louvinia Young Pleasant, in Rawick, *The American Slave*, Texas Narratives, supplement 2, vol. 8, pp. 3101–3102.

CHAPTER 4

1. B. D. Dashiell to William W. Browning, n.d., Browning Papers, DBCAH.

2. Rebecca Pilsbury Diary, November 18, 1848; December 23, 1848, SHC. The italics are mine.

3. Quoted in Mattie Austin Hatcher, *Letters of an Early American Traveller: Mary Austin Holley, Her Life and Works, 1784–1846* (Dallas: Southwest Press, 1933), 168.

4. Paul D. Lack, *The Texas Revolutionary Experience: A Political and Social History, 1835–1836* (College Station: Texas A&M Press, 1992), 240; William W. Freehling, *Prelude to Civil War: The Nullification Controversy in South Carolina, 1816–1836* (New York: Harper and Row, 1965); William W. Freehling, *The Road to Disunion: Secessionists at Bay, 1776–1854* (New York: Oxford University Press, 1990), 253–286, 308–352; McCurry, *Masters of Small Worlds*, 256–265.

5. Lack, *Texas Revolutionary Experience*, 15; J. G. McNeel to James F. Perry, June 22, 1835, in *AP*, 3:77; *Brazoria Texas Republican*, October 17, 1835. On the timing of the reconciliation between the War and Peace parties, see Lack, *Texas Revolutionary Experience*, 30–35; Cantrell, *Stephen F. Austin*, 306–316.

6. Quoted in William C. Binkley, *The Texas Revolution* (Baton Rouge: Louisiana State University Press, 1952), 60.

7. Guy M. Bryan to James F. Perry, December 13, 1844, Perry Papers, DBCAH.

8. Guy M. Bryan to Rutherford B. Hayes, November 18, 1847, in "The Bryan-Hayes Correspondence," ed. E. W. Winkler, *SHQ* 25 (October 1921): 115.

9. *A Visit to Texas in 1831*, 161.

10. Lack, *Diary of William Fairfax Gray*, 216–217.

11. *Anderson Texas Baptist*, February 27, 1856.

12. Ibid., April 25, 1855.

13. John Washington Lockhart Diary, 1859, Lockhart Papers, RL; U.S. Census Bureau, Slave Schedule, Washington County, Texas, 1860, record M653, roll 1312 (microfilm).

14. Oscar Addison Private Journal, June 2, 1861, Addison Papers, DBCAH.

15. Instructions for Road District 1, Undated General Correspondence, Pre-1865, Perry Papers, DBCAH.

16. Constitution of the Brazoria County Agricultural Society, Perry Papers, DBCAH. See also Subscription List for a Planters' Exchange, Groce Family Papers, DBCAH.

17. *Anderson Texas Baptist*, April 11, 1855.

18. Petition, June 27, 1852, Records of the Governor, Peter Hansbrough Bell, box 301-21, folder 34, TSL.

19. Gideon Lincecum to Matson, May 28, 1860, Letter Press Book, 1859–1860, Gideon Lincecum Collection, DBCAH.

20. Campbell, *An Empire for Slavery*, 100–101; *LT*, 2:1497–1498.

21. Rosa Kleberg, "Some of My Early Experiences in Texas, II," *TSHAQ* 1 (October 1898): 173.

22. Frederich Bay receipt, March 6, 1855, Lockhart Papers, RL.

23. "Josef Silar," in Machann and Mendl, *Czech Voices*, 29.

24. Olmsted, *A Journey through Texas*, 359.

25. Nagel, *A Boy's Civil War Story*, 117–118; Trenckmann, Autobiography, Trenckmann Papers, DBCAH, 3.

26. U.S. Census Bureau, Free Schedule, Washington County, 1860, record M653, roll 1307 (microfilm); A List of Church Members at Chappell Hill, Browning Papers, DBCAH; Winfield and Winfield, *All Our Yesterdays*, 31, 34.

27. Struve, *Germans and Texans*, 93, 136.

28. U.S. Congress, House Executive Documents, *Freedmen's Bureau. Letter from the Secretary of War, in Answer to a Resolution of the House of March 8, Transmitting a Report, by the Commissioner of the Freedmen's Bureau, of All Orders Issued by Him or Any Assistant Commissioner*, 39th Cong., 1st sess., 1866, nos. 70, 307.

29. Nagel, *A Boy's Civil War Story*, 40–41.

30. Ibid., 45.

31. On the subjectivity of ethnic identity, see Kathleen Neils Conzen, David A. Gerber, Ewa Morawska, George E. Pozzetta, and Rudolph J. Vecoli, "The Invention of Ethnicity: A Perspective from the U.S.A.," *Journal of American Ethnic History* 12 (Fall 1992): 3–41.

32. Nagel, *A Boy's Civil War Story*, 39; See also Trenckmann, Autobiography, Trenckmann Papers, DBCAH, 3.

33. Karl Urbantke, *Texas Is the Place for Me: The Autobiography of a German Immigrant Youth, Carl Urbantke*, trans. Ella Urbantke Fischer (Austin: Pemberton Press, 1970), 36. Urbantke converted to Methodism because there was no Lutheran Church. See also Craig William Borchardt, "German Lutheran Transplants and Methodist Converts in Washington and Austin Counties, Texas, 1860–1930" (Ph.D. diss., Texas A&M University, 1996).

34. Nagel, *A Boy's Civil War Story*, 54.

35. Ibid., 89.

36. "Jan Ustyník," in Machann and Mendl, *Czech Voices*, 43–44.

37. "Josef L. Lesikar," in Machann and Mendl, *Czech Voices*, 16.

38. Urbantke, *Texas Is the Place for Me*, 38.

39. Adalbert Regenbrecht, "The Settlers of Millheim before the Civil War," *SHQ* 20 (July

1916): 28–32. Several memoirs make the unverifiable (and unlikely) claim that many or most Germans in western Austin County came from professional backgrounds and the ranks of the "Forty-Eighter" republicans. See Kleberg, "Early Experiences in Texas, II," 173; Nagel, *A Boy's Civil War Story,* 40–44; Trenckmann, Autobiography, Trenckmann Papers, 7, DBCAH; Urbantke, *Texas Is the Place for Me,* 16; "Josef L. Lesikar," in Machann and Mendl, *Czech Voices,* 10.

40. Wilhelm Schlottmann to "Lieben Freunde," May, 1859, in Kammeier, *"Hallelujah,"* 48.

41. See entries for August 2, 1857; September 7, 1856, Agricultural Society of Cat Spring, Minutes, 1856–69, DBCAH.

42. Urbantke, *Texas Is the Place for Me,* 16.

43. See entries for August 30, 1857; May 30, 1858, Agricultural Society of Cat Spring, Minutes, 1856–69, DBCAH.

44. Nagel, *A Boy's Civil War Story,* 193.

45. Entry for June 21, 1858, Agricultural Society of Cat Spring, Minutes, 1856–69, DBCAH.

46. Election Returns, 1854, TSL. Returns for the other lower Brazos counties do not survive.

47. Sarah Ford, in Rawick, *The American Slave,* series 2, vol. 4, pt. 3, Texas Narratives, 1365–1366.

48. Texas County Tax Rolls, Austin County (1st series), reel 19; "Sketch of the Lives of Mary Loewenberger Amsler and Charles Conrad Amsler, by John C. Amsler," typescript, Amsler Papers, DBCAH.

49. Franz Hinze to "Freund Tödtenheide," February 25, 1862, in Kammeier, *"Hallelujah,"* 80.

50. Jordan, *German Seed in Texas Soil,* 110; Terry G. Jordan, "Germans and Blacks in Texas," in Miller, *States of Progress,* 92–94.

51. Cornelia Kuffner, "Texas Germans' Attitudes toward Slavery: Biedermeier Sentiments and Class-Consciousness in Austin, Colorado, and Fayette Counties" (master's thesis, University of Houston, 1994), 124–125.

52. Walter D. Kamphoefner, "New Perspectives on Texas Germans and the Confederacy," *SHQ* 102 (April 1999): 441–455; Walter D. Kamphoefner, "German Texans: In the Mainstream or Backwaters of Lone Star Society?" *Yearbook of German-American Studies* 38 (2003): 119–123.

53. Gustav Dresel, Texas Diary, 1837–1841, typescript, 69, DBCAH.

54. Urbantke, *Texas Is the Place for Me,* 13.

55. Kuffner, "Texas Germans' Attitudes toward Slavery," 4.

56. Kattner, "Diversity of Old South White Women," 299–309.

57. Franz Hinze to "Freund Tödtenheide," February 25, 1862, in Kammeier, *"Hallelujah,"* 80.

58. Viktor Bracht, *Texas in 1848,* trans. and ed. Charles Frank Schmidt (San Antonio: Naylor, 1931), 74.

59. Trenckmann, Autobiography, Trenckmann Papers, 26, DBCAH; Nagel, *A Boy's Civil War Story,* 93.

60. Friedrich Gerstäcker, "Die Backwoodsman Nordamerikas," in his *Ausgewählte Erzählungen und Humoresken* (Leipzig, Germany: Max Hesses Verlag, n.d.), 182.

61. Friedrich Gerstäcker, *The Wanderings and Fortunes of Some German Immigrants,* trans. David Black (London: D. Bogue, 1848).

62. Charles Sealsfield, *Der Pflanzer am Red River* (Berlin: K. Siegismund, 1943); Jerry

Schuchalter, *Frontier and Utopia in the Fiction of Charles Sealsfield: A Study of the Lebensbilder aus der westlichen Hemisphäre* (Frankfurt: Peter Lang, 1986), 188–213.

63. Shearer Davis Bowman, *Masters and Lords: Mid-Nineteenth Century U.S. Planters and Prussian Junkers* (New York: Oxford University Press, 1993), 22–23; Rainer Koch, "Liberalismus, Konservatismus, und das Problem der Negersklaverei: Ein Beitrag zur Geschichte der politischen Denkens in Deutschland in der ersten Hälfte des 19. Jahrhunderts," *Historische Zeitschrift* 222 (June 1976): 529–577.

64. Nagel, *A Boy's Civil War Story*, 115.

65. Urbantke, *Texas Is the Place for Me*, 17.

66. Randolph B. Campbell, *Gone to Texas: A History of the Lone Star State* (New York: Oxford University Press, 2003), 174–181. On the Brazoria County participants, see Creighton, *A Narrative History of Brazoria County*, 179–181.

67. Election Returns, 1851, 1852, 1853, 1855, 1856, 1857, 1859, TSL. In 1855, Grimes County gave Dickson 93 percent of the vote, but that was almost certainly because he lived in Anderson.

68. *Galveston Weekly News*, March 6, 1855.

69. *Anderson Central Texian*, January 1, 1855.

70. Earl Wesley Fornell, *The Galveston Era: The Texas Crescent on the Eve of Secession* (Austin: University of Texas Press, 1961), 170–179; Suzanne Lynn Summers, "The Role of Merchants in Promoting Economic Development in the Houston-Galveston Region, 1836–1860" (Ph.D. diss., University of Texas, 1996), 281–361.

71. Gregg Cantrell, "Sam Houston and the Know-Nothings: A Reappraisal," *SHQ* 96 (January 1993): 326–343.

72. *Washington American*, November 1, 1855.

73. Ibid., November 8, 1855.

74. David Roediger, *The Wages of Whiteness: Race and the Making of the American Working Class* (London: Verso Press, 1991), 155.

75. *Washington American*, November 30, 1855.

76. Ibid., April 23, 1856.

77. Campbell, *An Empire for Slavery*, 224–228; Walter L. Buenger, *Secession and the Union in Texas* (Austin: University of Texas Press, 1984), 55–59.

78. Buenger, *Secession and the Union in Texas*, 49; quote in Henry Steele Commager, ed., *Documents of American History*, 6th ed. (New York: Appleton, Century, Crofts, 1958), 366.

79. James McKenzie to Hugh C. Cooper, September 18, 1860, Hugh C. Cooper Papers, SHC.

80. Kamphoefner, "New Perspectives," 441–455.

81. Declaration, November 17, 1860, Mordello Steven Munson Personal Correspondence, Munson Family Papers, DBCAH.

82. Buenger, *Secession and the Union in Texas*, 125.

83. *LT*, 5:1519.

84. Trenckmann, Autobiography, Trenckmann Papers, 14, DBCAH.

85. Creighton, *A Narrative History of Brazoria County*, 230.

86. Dale Baum, *The Shattering of Texas Unionism: Politics in the Lone Star State during the Civil War Era* (Baton Rouge: Louisiana State University, 1998), 60–63, 72, uses regression analysis to

demonstrate that in the February 1861 secession referendum, Grimes, Brazoria, Austin, and Fort Bend counties all favored secession at rates significantly higher than might be predicted based on a comparison with ballots cast for John C. Breckinridge in the November 1860 election. The reasons why are not clear. Baum's suggestion that secessionist "exhilaration and resolve" accounts for their overwhelming success in these and neighboring counties is certainly correct, but it is also important to recognize that "exhilaration and resolve" can intimidate the opposition and stifle dissent.

87. Entry for March 6, 1861, Oscar Addison Private Journals, Addison Papers, DBCAH.

CHAPTER 5

1. Thomas Blackshear Diary and Memorandum Book, 1861–1862, Blackshear Papers, DBCAH.

2. U.S. Census Bureau, *Population of the United States in 1860*, 484–487; U.S. Census Bureau, *Statistical View of the United States*, 308–315.

3. Entry for March 3, 1861, Oscar Addison Private Journals, Addison Papers, DBCAH.

4. Entry for March 6, 1861, ibid.

5. Blackshear Diary and Memorandum Book, vol. 3, Blackshear Papers, DBCAH; Peach Point Plantation Ledger, 48–63, Perry Papers, DBCAH. For a complete discussion of planters' efforts to create dependent slave patriarchies, see chapter 3.

6. Abigail Curlee Holbrook, "Cotton Marketing in Antebellum Texas," *SHQ* 63 (April 1970): 431–455.

7. On Confederate diplomacy, see Frank Lawrence Owsley, *King Cotton Diplomacy: Foreign Relations of the Confederate States of America*, 2nd ed., rev. by Harriet Owsley (Chicago: University of Chicago Press, 1959).

8. Cat Spring Agricultural Society, *The Cat Spring Story* (San Antonio: Lone Star, 1956), 136, 144.

9. McMahan and Gilbert to John Washington Lockhart, June 6, 1862, Lockhart Papers, RL. See also Ronnie C. Tyler, "Cotton on the Border," *SHQ* 63 (April 1970).

10. Cora Bryan to Sarah Perry, September 5, 1863, Perry Papers, DBCAH.

11. Promissory Notes dated December 19, 1863; February 29, April 9, May 1, October 10, 1864, Perry Papers, DBCAH.

12. On the role of impressment in eroding slavery, see Joseph P. Reidy, *From Slavery to Agrarian Capitalism in the Cotton Plantation South: Central Georgia, 1800–1880* (Chapel Hill: University of North Carolina Press, 1992), 130; James Roark, *Masters without Slaves: Southern Planters in the Civil War and Reconstruction* (New York: Norton, 1977), 78–81; Campbell, *An Empire for Slavery*, 234–239.

13. Thomas Affleck to Enrolling Officer, Washington County, Texas, October 17, 1864, box X-59, letter book 3, finished ribbon copy, 648, Affleck Papers, HML.

14. John Washington Lockhart to William W. Browning, November 1, 1863, Browning Papers, DBCAH.

15. For evidence of planter concern about slave discipline in the army, see Reidy, *From Slavery to Agrarian Capitalism*, 131; Armstead Robinson, "Day of Jubilo: Civil War and the Demise of

Slavery in the Mississippi Valley, 1861–1865" (Ph.D. diss, University of Rochester, 1976), 179–183; Roark, *Masters without Slaves*, 80–81.

16. CSA Promissory Notes to Stephen S. Perry, Correspondence, Perry Papers, DBCAH; slave family data is based on a list of slave couples and children in the Peach Point Plantation Ledger, Perry Papers, DBCAH.

17. James F. Perry Estate Inventory, 1853, Perry Papers, DBCAH, gives the slaves' ages.

18. Lizzie Scott Neblett to Will Neblett, August 5, 1863, Neblett Papers, DBCAH.

19. Hiram Brown to Sarah Perry, July 2, 1864, Perry Papers, DBCAH.

20. Will Neblett to Lizzie Scott Neblett, July 28, 1863; Lizzie Scott Neblett to Will Neblett, August 5, 1863, Neblett Papers, DBCAH.

21. F. G. Brown to William W. Browning, January 3, 1864, Browning Papers, DBCAH.

22. *Bellville Countryman*, May 23, 1863; Louis Lehmann to Frederike Lehmann, December 19, 1863, German Immigrant Letters, LOC.

23. Wood to John Washington Lockhart, November 11, 1863, Browning Papers, DBCAH. The law exempted one military-age man for every twenty slaves in a household. Wood probably did not receive an exemption for his work at Lockhart's because the exemptions were taken by Lockhart and his family members.

24. Ibid.

25. Ibid.

26. Ibid.

27. John Washington Lockhart to William W. Browning, November 1, 1863, Browning Papers, DBCAH.

28. See chapter 4.

29. Randy J. Sparks, "John P. Osterhout, Yankee, Rebel, Republican," *SHQ* 90 (October 1986), 116–117.

30. "Jan Lesikar" in Machann and Mendl, *Czech Voices*, 18. For Engelking's political stance, see *Bellville Countryman*, August 21, 1861; for Kleberg and Langhammer, see Trenckmann, Autobiography, Trenckmann Papers, 14–16, DBCAH.

31. Nagel, *A Boy's Civil War Story*, 207–208. See also Trenckmann, Autobiography, Trenckmann Papers, 15, DBCAH.

32. "Jan Ustiník," in Machann and Mendl, *Czech Voices*, 46, 51–62.

33. F. A. Thompson to Stephen S. Perry, November 4, 1861, Perry Papers, DBCAH.

34. *Bellville Countryman*, January 3, 1863.

35. Friedrich Winkelmann to Geliebte Eltern, n.d., 1863, in Kammeier, *Ich muß mir Ärgern*, 178; Nagel, *A Boy's Civil War Story*, 209–210.

36. J. P. Flewellen to Edmund P. Turner, December 4, 1862, in *OR*, series 1, 15:886; A. J. Bell to Maj. J. P. Flewellen, January 3, 1863, ibid., 925–926; William G. Webb to Maj. A. G. Dickinson, January 4, 1863, ibid., 926–928.

37. *Houston Tri-Weekly Telegraph*, February 27, 1863.

38. Henry L. Webb to Edmund P. Turner, February 18, 1863, in OR, series 1, 15:981–982; "Report of William J. Wheeler concerning Certain Charges Preferred against Lieut. R. H. Stone by Certain German Citizens of New Ulm and its Vicinity," ibid., 15:989.

39. *Houston Tri-Weekly Telegraph*, February 27, 1863.

40. Ibid., March 11, 1863; *Bellville Countryman*, March 7, 1863. The article changed "McElroy" to "Muckleroy."

41. *Bellville Countryman*, April 4, 1863.

42. The list began running in the *Bellville Countryman* on December 20, 1862, and continued into the first weeks of the following year. Almost all came from Cat Spring, New Ulm, and Industry, the three largest German settlements in Austin County.

43. "Jan Lesikar," in Machann and Mendl, *Czech Voices*, 18, says there were "many" lynchings and implies that both Anglo and German Confederates participated.

44. Eugene D. Genovese, *From Rebellion to Revolution: Afro-American Slave Revolts in the Making of the Modern World* (Baton Rouge: Louisiana State University Press, 1979), argues that slave rebellions after Haiti were bourgeois in nature. Michael Craton, *Testing the Chains: The Resistance to Slavery in the British West Indies* (Ithaca, NY: Cornell University Press, 1982), argues that European ideas had minimal influence on slave resistance. For the French Caribbean, see Laurent Dubois, *A Colony of Citizens: Revolution and Slave Emancipation in the French Caribbean, 1787–1804* (Williamsburg, VA, and Chapel Hill: University of North Carolina Press for the Omohundro Institute of Early American Culture, 2004); Laurent Dubois, *Avengers of the New World: The Story of the Haitian Revolution* (Cambridge, MA: Belknap Press of Harvard University, 2004); for Jamaica, see Thomas Holt, *The Problem of Freedom: Race, Labor, and Politics in Jamaica and Britain, 1832–1938* (Baltimore: Johns Hopkins University Press, 1992). For Cuba, see Rebecca J. Scott, *Slave Emancipation in Cuba: The Transition to Free Labor, 1865–1899* (Princeton, NJ: Princeton University Press, 1986).

45. Sallie Munson to Mordello Stephen Munson, January 12, 1863, Munson Family Papers, DBCAH.

46. Campbell, *An Empire for Slavery*, 246–248, argues that Texas slaves knew of the war and its significance for the end of slavery but did little to interfere with the war effort.

47. *Bellville Countryman*, September 5, 1863, 2.

48. Lizzie Scott Neblett to Will Neblett, November 13, 1863; Will Neblett to Lizzie Scott Neblett, November 22, 1863, Neblett Papers, DBCAH. For a discussion of race and sex during the war, see Martha Hodes, "Wartime Dialogues on Illicit Sex: White Women and Black Men," in *Divided Houses: Gender and the Civil War*, ed. Catherine Clinton and Nina Silber (New York: Oxford University Press, 1992): 230–242.

49. Lizzie Scott Neblett to Will Neblett, August 11, 1863, Neblett Papers, DBCAH; *Bellville Countryman*, October 17, 1863.

50. Drew Gilpin Faust, *Mothers of Invention: Women of the Slaveholding South in the American Civil War* (New York: Vintage Books, 1996), 65–70, offers an insightful interpretation of the same incident. While I generally agree with her interpretation, I have chosen to stress slightly different themes.

51. *Bellville Countryman*, October 24, 1863.

52. Lizzie Scott Neblett to Will Neblett, August 5, 1863; Lizzie Scott Neblett to Will Neblett, August 11, 1863, Neblett Papers, DBCAH.

53. Lizzie Scott Neblett to Will Neblett, August 11, 1863, ibid.

54. Lizzie Scott Neblett to Will Neblett, October 25, 1863, ibid.

55. Lizzie Scott Neblett to Will Neblett, November 23, 1863, ibid.

56. Ibid.

57. Ibid.

58. Ibid.

59. Lizzie Scott Neblett to Will Neblett, November 25, 1863, ibid.

60. Ibid.

EPILOGUE

1. *Brooklyn Daily Eagle*, June 30, 1889, 9.

2. For the text of Granger's order, see *OR*, vol. 48, pt. 2, p. 929. On the planter transition from "labor lords" to "landlords," see Gavin Wright, *Old South, New South: Revolutions in the Southern Economy since the Civil War* (New York: Basic Books, 1986), 17–50. My understanding of emancipation is shaped significantly by the work of the Freedmen and Southern Society Project. For a brief summary of those interpretations, see Ira Berlin, Barbara J. Fields, Steven F. Miller, Joseph P. Reidy, and Leslie S. Rowland, *Slaves No More: Three Essays on Emancipation and the Civil War* (Cambridge: Cambridge University Press, 1992. Other influential works include Eric Foner, *Reconstruction: America's Unfinished Revolution, 1863–1877* (New York: Harper and Rowe, 1988); Holt, *The Problem of Freedom;* Rebecca J. Scott, *Degrees of Freedom: Louisiana and Cuba after Slavery* (Cambridge, MA: Belknap Press of Harvard University Press, 2005). For a more detailed discussion of emancipation along the Brazos, see Sean M. Kelley, "Plantation Frontiers: Race, Ethnicity, and Family along the Brazos River of Texas, 1821–1886" (Ph.D. diss., University of Texas, 2000).

3. U.S. Congress, House Executive Documents, *Freedmen's Bureau*, nos. 70, 375, 309, 305.

4. Kelley, "Plantation Frontiers," chap. 7. The Freedmen's Bureau officially concluded operations on January 1, 1869, although the schools continued into 1870. See William L. Richter, *Overreached on all Sides: The Freedmen's Bureau Administrators in Texas, 1865–1868* (College Station: Texas A&M University Press, 1991), 287.

5. On the smallholders of Brazoria County, see Sean Kelley, "A Texas Peasantry? Black Smallholders in the Texas Sugar Bowl, 1865–1890," *Slavery and Abolition* 28 (August 2007): 193–209.

6. For an overview of Reconstruction in Texas, see Carl H. Moneyhon, *Texas after the Civil War: The Struggle for Reconstruction* (College Station: Texas A&M University Press, 2004); James Smallwood, Barry A. Crouch, and Larry Peacock, *Murder and Mayhem: The War of Reconstruction in Texas* (College Station: Texas A&M University Press, 2003).

7. Kelley, "Plantation Frontiers," chap. 8.

8. U.S. Congress, Senate, Miscellaneous Documents, *Testimony on the Alleged Election Outrages in Texas Reported from the Committee on Privileges and Elections of the Senate of the United States*, 50th Cong., 2nd sess., 1889. See also Donald G. Nieman, "Black Political Power and Criminal Justice: Washington County, Texas, 1868–1884," *Journal of Southern History* 55 (August 1989): 391–420; Donald G. Nieman, "African Americans and the Meaning of Freedom: Washington County, Texas, as a Case Study, 1865–1886," *Chicago-Kent Law Review* 70, no. 2 (1994):

541–582; Donald G. Nieman, "African American Communities, Politics, and Justice: Washington County, Texas, 1865–1890," in *Local Matters: Race, Crime, and Justice in the Nineteenth-Century South*, ed. Christopher Waldrep and Donald G. Nieman (Athens: University of Georgia Press, 2001), 201–224; Robert W. Shook, "'The Texas Election Outrage' of 1886," *East Texas Historical Journal* 10 (Spring 1972); Kelley, "Plantation Frontiers," chap. 8.

9. Leslie A. Lovett, "Biracial Politics and Community Development: The Reconstruction Experience in Fort Bend County, Texas, 1869–1889," *Houston Review* 16 (Fall 1994): 27–39; Alwyn Barr, "Reconstruction Change and Continuity in Brazoria County, Texas," *Houston Review* 18 (Spring 1996): 120–121; Lawrence Goodwyn, "Populist Dreams and Negro Rights: East Texas as a Case Study," *American Historical Review* 76 (December 1971): 1135–1171.

10. J. P. Underwood Scrapbook, BCHM.

11. Foley, *White Scourge*, 40–63.

12. U.S. Census Fact Sheets, http://factfinder.census.gov, accessed August 29, 2006.

13. *Brazosport Facts*, March 11, 2006.

Bibliography

PRIMARY SOURCES

Unpublished

Austin County Clerk, Bellville, Texas (ACC)
- Colonial Records, 1810–37, 2 vols.
- Minutes of the Commissioners' Court
- Probate Files

Baker Library Historical Collections, Harvard University School
- R. G. Dun and Company Collection

Bancroft Library, University of California, Berkeley
- Herbert Howe Bancroft Collection, Texas Narratives

Brazoria County Clerk, Angleton, Texas (BCC)
- Deeds
- Probate Files

Brazoria County Historical Museum, Angleton, Texas (BCHM)
- Ammon Underwood Scrapbook
- Records of the District Court of Brazoria County (DCBC)

Dolph Briscoe Center for American History, University of Texas at Austin (DBCAH)
- Oscar Addison Papers
- John Adriance Papers
- Agricultural Society of Cat Spring Minutes
- American Missionary Association Manuscripts, microfilm ed. (originals available at Amistad Research Center, Tulane University, New Orleans)
- John C. Amsler Papers
- Eugene C. Barker Papers
- William Bartels Papers
- James H. Bell Papers
- Black History Collection
- Thomas Edward Blackshear Papers
- William Bradbury Papers
- William W. Browning Papers
- Guy M. Bryan Papers
- Nestor and Tacitus Clay Papers
- John P. Coles Papers
- David C. Dickson Papers

Thomas Dwyer Papers
Hermann Ernst Account Book
Charles W. Fordtran Papers
William Fairfax Gray Diaries
Groce Family Papers
Anson Jones Papers
Julius Lehmann Papers
Leurson-Franke Papers
Gideon Lincecum Papers
Munson Family Papers
Lizzie Scott Neblett Papers
James F. and Stephen S. Perry Papers
William A. Trenckmann Papers
Ammon Underwood Papers
William P. Zuber Papers

Chappell Hill Historical Society, Chappell Hill, TX
Miscellaneous archival collections

District Court of Austin County, Bellville, Texas (DCAC)
Case Files
Civil Court Minutes

District Court of Washington County, Brenham, Texas (DCWC)
Final Record of the District Court of Washington County
Records of the District Court of Washington County

Freedmen and Southern Society Project, University of Maryland
Miscellaneous files

General Land Office, Austin, Texas (GLO)
Austin Colony Census
County Survey Maps
Guide to Spanish and Mexican Land Grants in Texas

Hill Memorial Library, Louisiana State University, Baton Rouge (HML)
Thomas Affleck Papers
Claiborne Manuscripts
John H. Randolph Papers

Library of Congress, Washington, DC (LOC)
American Life Histories
Auswandererbriefe [German Immigrant Letters]
Lewis Tappan Papers

National Archives and Records Administration, Washington, DC
Despatches from U.S. Consuls in Havana, Cuba (microfilm), RG 59
Records of the Bureau of Refugees, Freedmen, and Abandoned Lands, RG 105

Private Collection
Levi Jordan Papers

Rosenberg Library, Galveston (RL)
Thomas Affleck Papers
John Washington Lockhart Papers
Wallis Family Papers
Samuel May Williams Papers
Southern History Collection, University of North Carolina, Chapel Hill (SHC)
Hugh C. Cooper Papers
Ernest Haywood Papers
Theodore Davidson Morrison Papers
Robert Treat Paine Papers
Rebecca Pilsbury Diary
Texas State Library, Archive Division, Austin (TSL)
Convention Records, 1836
Election Results
Governors' Papers
Texas County Tax Rolls (microfilm)
Washington County Clerk, Brenham, Texas (WCC)
Deeds
Probate Files
Transcript of the County Court Record

Published

GOVERNMENT DOCUMENTS AND ORGANIZATIONAL REPORTS

Howard, Benjamin C., comp. Reports of Cases Argued and Determined in the Supreme Court of the United States, 1850–1851. 64 vols. Washington, DC: John Conrad and Co., 1804–61.

U.S. Census Bureau. *Agriculture of the United States in 1860; Compiled from the Original Returns of the Eighth Census.* Washington, DC: Government Printing Office, 1864.

———. *Population of the United States in 1860, Compiled from the Original Returns of the Eighth Census.* Washington, DC: Government Printing Office, 1864.

———. *The Seventh Census of the United States: 1850.* Washington, DC: Robert Armstrong, 1853.

———. *Statistical View of the United States.* Washington, DC: A. O. P. Nicholson, 1854.

———. *Statistics of the United States (Including Mortality, Property, &c.) in 1860.* Washington, DC: Government Printing Office, 1866.

U.S. Congress, House Executive Documents. *Freedmen's Bureau. Letter from the Secretary of War, in Answer to a Resolution of the House of March 8, Transmitting a Report, by the Commissioner of the Freedmen's Bureau, of All Orders Issued by Him or Any Assistant Commissioner.* 39th Cong., 1st sess., 1866.

U.S. Congress, Senate. *Slaves Imported by an Indian Agent Contrary to Law.* 17th Cong., 1st sess., 1822.

U.S. Congress, Senate, Committee on Privileges and Elections. *Alleged Outrages in Washington County, Texas.* Washington, DC: Government Printing Office, 1889.

U.S. Congress, Senate, Miscellaneous Documents. *Testimony on the Alleged Outrages in Texas Reported from the Committee on Privileges and Elections of the Senate of the United States.* 50th Cong., 2nd sess., 1889.

U.S. State Department. *Despatches from U.S. Consuls in Havana.* Copy No. T-20, National Archives and Records Administration Microfilm.

U.S. War Department. *The War of the Rebellion: A Compilation of the Official Records of the Union and Confederate Armies.* Edited by Robert N. Scott. 128 vols. Washington, DC: Government Printing Office, 1880–1901. (*OR*)

BOOKS

Affleck, Thomas. *Affleck's Southern Rural Almanac.* Brenham, TX: N.p., 1860; Brenham, TX: New Year's Creek Settlers' Association, 1986.

Baldwin, Joseph G. *The Flush Times of Alabama and Mississippi: A Series of Sketches.* With an introduction by James H. Justus. Baton Rouge: Louisiana State University Press, 1987.

Barker, Eugene C., ed. *The Austin Papers.* 3 vols. Washington, DC: Government Printing Office, 1924–28. (*AP*)

Berlandier, Jean Louis, *The Indians of Texas in 1830.* Edited by John C. Evers. Washington, DC: Smithsonian Institution Press, 1969.

Bolton, Herbert Eugene. *The Spanish Borderlands: A Chronicle of Old Florida and the Southwest.* New Haven, CT: Yale University Press, 1921.

Bouchereau, Louis. *Statement of the Sugar and Rice Crops Made in Louisiana, 1870–71.* New Orleans: Young, Bright, 1870.

Bracht, Viktor. *Texas in 1848.* Translated and edited by Charles Frank Schmidt. San Antonio: Naylor, 1931.

Braman, D. E. E. *Braman's Information about Texas.* Philadelphia: Lippincott, 1858.

Breeden, James O., ed. *Advice among Masters: The Ideal of Slave Management in the Old South.* Westport, CT: Greenwood Press, 1980.

Burns, Robert I., S.J., ed. *Las Siete Partidas.* 5 vols. Translated by Samuel Parsons Scott. Philadelphia: University of Pennsylvania Press, 2001.

Cat Spring Agricultural Society. *Century of Agricultural Progress, 1856–1956: Minutes of the Cat Spring Agricultural Society.* San Antonio: Lone Star, 1956.

Catterall, Helen Tunnicliff, ed. *Judicial Cases concerning American Slavery and the Negro.* 5 vols. New York: Octagon Books, 1968.

Celebrated and Extraordinary Trial of Col. Monroe Edwards for Forgery and Swindling. New York: Herald Office, 1842; Austin: Pemberton Press, 1970.

Champomier, P. A. *Statement of the Sugar Crop Made in Louisiana.* New Orleans: Cook, Young, 1852–58.

Commager, Henry Steele, ed. *Documents of American History.* 6th ed. New York: Appleton, Century, Crofts, 1958.

Conrad, Robert Edgar, ed. *Children of God's Fire: A Documentary History of Black Slavery in Brazil.* Princeton, NJ: Princeton University Press, 1983.

Curtin, Philip D., ed. *Africa Remembered: Narratives by West Africans in the Era of the Slave Trade.* Madison: University of Wisconsin Press, 1967.

De Bow, J. D. B. *The Industrial Resources of the Southern and Western States.* New Orleans: De Bow's Review, 1853.

DeWees, William B. *Letters from an Early Settler of Texas.* Louisville: Hull and Brother, 1853.

Dun, R. G., and Company. *Mercantile Agency Reference Book and Key.* New York: R. G. Dun, 1873–89.

Dunt, Detlef. *Reise Nach Texas nebst Nachrichten von diesem Lande: Für Deutsche, welche nach Amerika zu gehen beabsichtigen.* Bremen, Germany: Carl Wilh. Wiehe, 1834.

Exley, Jo Ella Powell. *Texas Tears and Texas Sunshine: Voices of Frontier Women.* College Station: Texas A&M University Press, 1985.

Farley, Hilde Shuptrine, and Alfred Earl Farley Jr. *Lou Ella Shuptrine's Letters to George William Harrison in Houston County, Texas, 1893–1895.* Shawnee Mission, KS: Privately published, 1996.

Faust, Drew Gilpin, ed. *The Ideology of Slavery: Proslavery Thought in the Antebellum South, 1830–1860.* Baton Rouge: Louisiana State University Press, 1981.

Foner, Philip S., and George E. Walker, eds. *Proceedings of the Black National and State Conventions, 1865–1900.* Philadelphia: Temple University Press, 1986.

Ford, John Salmon. *Rip Ford's Texas.* Edited and with an introduction and commentary by Stephen B. Oates. Austin: University of Texas Press, 1963.

Gammel, H. P. N., comp. *The Laws of Texas, 1822–1897.* 10 vols. Austin: Gammel, 1898. (*LT*)

Gates, Henry Louis, Jr., ed. *The Classic Slave Narratives.* New York: Penguin, 1987.

Gerstäcker, Friedrich. *Ausgewählte Erzählungen und Humoresken.* Leipzig, Germany: Max Hesses Verlag, n.d.

———. *The Wanderings and Fortunes of Some German Immigrants.* Translated by David Black. London: D. Bogue, 1848.

Goeth, Ottilie. *Was Grossmutter Erzaehlt.* San Antonio: Passing Show, 1915.

Gray, William Fairfax. *From Virginia to Texas: The Diary of William Fairfax Gray.* Houston: Fletcher Young, 1965.

Hatcher, Mattie Austin. *Letters of an Early American Traveller: Mary Austin Holley, Her Life and Works, 1784–1846.* Dallas: Southwest Press, 1933.

Holley, Mary Austin. *Texas.* Austin: Steck, 1935.

Hollon, W. Eugene, ed. *William Bollaert's Texas.* Foreword by Joe B. Frantz. Norman: University of Oklahoma Press, 1989.

Jenkins, John H., ed. *Papers of the Texas Revolution, 1835–1836.* Austin: Presidial Press, 1973.

Kammeier, Heinz-Ulrich, ed. *"Hallelujah, jetzt sehen wir Amerika": Auswandererbriefe aus dem Kreis Lübbecke und Umgebung, 1836–1889.* Espelkamp, Germany: Verlag Marie L. Leidorf, 1994.

———, ed. *"Ich muß mir ärgern, das ich nicht ehr übern Großen Ozean gegangen bin": Auswanderer aus dem Kreis Lübbecke und Umgebung berichten aus Amerika.* Espelkamp, Germany: Mittwalddruck, 1988.

———, ed. *"So besint euch doch nicht lange und kommt herrüber . . .": Briefe von Amerikaauswanderern aus dem Kreis Lübbecke aus zwei Jahrhunderten.* Darmstadt, Germany: Dissertations Druck, 1989.

Kamphoefner, Walter D., and Wolfgang Helbich., eds. *Germans in the Civil War: The Letters They Wrote Home.* Translated by Susan Carter Vogel. Chapel Hill: University of North Carolina Press, 2006.

Kamphoefner, Walter D., Wolfgang Helbich, and Ulrike Sommer, eds. *News from the Land of Freedom: German Immigrants Write Home.* Ithaca, NY: Cornell University Press, 1991.

King, Richard, ed. *A Victorian Lady on the Texas Frontier: The Journal of Ann Raney Coleman.* Norman: University of Oklahoma Press, 1971.

Lack, Paul D., ed. *The Diary of William Fairfax Gray: From Virginia to Texas, 1835–1837.* Dallas: Southern Methodist University Press, 1997.

Law, Robin, and Paul Lovejoy, eds. *The Biography of Mahommah Gardo Baquaqua: His Passage from Slavery to Freedom in Africa and America.* Princeton, NJ: Markus Wiener, 2001.

Life and Adventures of the Accomplished Forger and Swindler, Colonel Monroe Edwards. Philadelphia: T. B. Peterson and Brothers, 1848.

Life of the Celebrated Munroe Edwards. Boston: William White and H. P. Lewis, 1842.

Lincecum, Jerry Bryan, Edward H. Phillips, and Peggy A. Redshaw, eds. *Gideon Lincecum's Sword: Civil War Letters from the Texas Home Front.* Denton: University of North Texas Press, 2001.

Luckey, John. *Life in Sing Sing Prison, as Seen in a Twelve Years Chaplaincy.* New York: N. Tibbals, 1860.

Lundy, Benjamin. *Life, Travels, and Opinions of Benjamin Lundy.* Compiled by Thomas Earle. New York: Arno Press, 1969.

Machann, Clinton, and James W. Mendl Jr., eds. and translators. *Czech Voices: Stories from Texas in the Amerikan Narodni Kalendar.* College Station: Texas A&M University Press, 1991.

Murr, Erika L., ed. *A Rebel Wife in Texas: The Diary and Letters of Elizabeth Scott Neblett, 1852–1864.* Baton Rouge: Louisiana State University Press, 2001.

Nagel, Charles. *A Boy's Civil War Story.* St. Louis: Dorrance, 1937.

Northup, Solomon. *Twelve Years a Slave.* Auburn, NY: Derby and Miller, 1854; New York: Dover, 1970.

Olmsted, Frederick Law. *A Journey through Texas; or, A Saddle Trip on the Southwestern Frontier.* Austin: University of Texas Press, 1978.

Orenseth, Fisher. *Texas in 1840.* Waco, TX: Texian Press, 1964. First published in 1841.

Rawick, George P., ed. *The American Slave: A Composite Autobiography.* 19 vols.; supplement series 1, 12 vols.; supplement series 2, 10 vols. Westport, CT: Greenwood Press, 1972–79.

Schlecht, Friedrich. *Mein Ausflug nach Texas.* Bellville, TX: Bellville Wochenblatt, 1930.

———. *On to Texas! The Story of a Journey to Texas in 1848.* Including a biography of the author by Clara Matthaei. English translation and notes by Charles Patrick. Manor, TX: Indio Bravo Press, 1998.

Sealsfield, Charles [Karl Postl]. *Der Pflanzer am Red River.* Berlin: K. Siegismund, 1943.

Smithwick, Noah. *The Evolution of a State; or, Recollections of Old Texas Days.* Austin: University of Texas Press, 1983.

Stampp, Kenneth, ed., *Records of the Ante-Bellum Southern Plantations from the Revolution through the Civil War.* Series 1. Selections from Louisiana State University (microfilm), pt. 1, reel 14.

Sweet, Alex. E., and J. Armoy Knox. *On a Mexican Mustang through Texas, from the Gulf to the Rio Grande.* Hartford, CT: S. S. Scranton, 1883.

Taylor, Nathaniel. *The Coming Empire; or, Two Thousand Miles in Texas on Horseback.* Dallas: Turner, 1936.

Texas Almanac for the Year 1857. Galveston: Richardson, 1857.

Texian. *The Life of the Celebrated Munroe Edwards.* Boston: William White and H. P. Lewis, 1842.

Trenckmann, William. *Austin County: Beilage zum Bellville Wochenblatt.* Bellville, TX: Bellville Wochenblatt, ca. 1899.

Urbantke, Carl. *Texas Is the Place for Me: The Autobiography of a German Immigrant Youth, Carl Urbantke.* Translated by Ella Urbantke Fischer. Austin: Pemberton Press, 1970.

A Visit to Texas in 1831, Being the Journal of a Traveller through Those Parts Most Interesting to American Settlers. New York: Goodrich and Wiley, 1834; Reprint, Austin: Steck, 1952.

Wallis, Johnnie Lockhart, and L. L. Hill, eds. *Sixty Years on the Brazos: The Life and Letters of Dr. John Washington Lockhart, 1824–1900.* New York: Argonaut Press, 1966.

Williams, Charles Richard, ed. *Diary and Letters of Rutherford B. Hayes, the Nineteenth President of the United States,* vol. 1, *1834–1860.* Columbus: Ohio State Archaeological and Historical Association, 1922.

Williams, Stephen K., ed. *Cases Argued and Decided in the Supreme Court of the United States, 1850–1851.* Rochester, NY: Lawyers Co-operative, 1901.

Woods, Gary D., comp. *The Hicks-Adams-Bass-Floyd-Patillo and Collateral Lines, Together with Family Letters, 1840–1868.* Salado, TX: Anson Jones Press, 1963.

ARTICLES

Adams, Ephraim Douglas, ed. "British Correspondence concerning Texas." *Southwestern Historical Quarterly* 17 (October 1913): 188–206.

Almonte, Juan N. "Statistical Report on Texas." Translated by C. E. Castañeda. *SHQ* 28 (January 1925): 177–222.

Austin, Stephen F. "Journal of Stephen F. Austin on His First Trip to Texas, 1821." *Texas Historical Association Quarterly* 4 (April 1904): 286–307.

Burnam, Jesse. "Reminiscences of Capt. Jesse Burnam." *Texas State Historical Association Quarterly* 5 (July 1901): 12–18.

Harris, Dilue. "Reminisces of Mrs. Dilue Harris." *Quarterly of the Texas State Historical Association* 4 (April 1900): 85–127.

Hinueber, Caroline von. "Life of German Pioneers in Early Texas." *Texas State Historical Association Quarterly* 2 (January 1899): 227–232.

Holland, J. K. "Reminisces of Austin and Old Washington." *Texas Historical Association Quarterly* 1 (October 1897): 92–95.

Jordan, Gilbert, ed. and trans. "William Steinert's View of Texas in 1849." *Southwestern Historical Quarterly* 80 (July 1976): 57–78; 80 (October 1976): 177–200; 80 (January 1977): 283–301; 80 (April 1977): 399–416; 81 (July 1977): 45–72.

Kleberg, Rosa. "Some of My Early Experiences in Texas." *Texas State Historical Association Quarterly* 1 (April 1898): 297–302.

———. "Some of My Early Experiences in Texas, II." *Texas State Historical Association Quarterly* 1 (October 1898): 170–173.

Kuykendall, J. H., ed. "Reminiscences of Early Texans: Capt. Gibson Kuykendall." *Texas State Historical Association Quarterly* 7 (July 1903): 29–39.

———. "Reminiscences of Early Texans: A Collection from the Austin Papers." *Texas State Historical Association Quarterly* 6 (January 1903): 236–241.

Lewis, W. S. "Adventures of the 'Lively' Immigrants, I." *Quarterly of the Texas State Historical Association* 3 (July 1899): 1–32.

———. "Adventures of the 'Lively' Immigrants, II." *Quarterly of the Texas State Historical Association* 3 (October 1899): 81–107.

Regenbrecht, Adalbert. "The German Settlers of Millheim before the Civil War." *Southwestern Historical Quarterly* 20 (July 1916): 28–34.

Rogers, William Warren, ed. "From Planter to Farmer: A Georgia Man in Reconstruction Texas." *Southwestern Historical Quarterly* 72 (April 1969): 526–529.

Sánchez, José María. "A Trip to Texas in 1828." Translated by Carlos E. Castaneda. *Southwestern Historical Quarterly* 29 (April 1926): 249–288.

Stoehr, Louise Ernst. "Die Erste Deutsche Frau in Texas." *Deutsche Pionier* 16 (1884): 372–375.

Watters, Jonathan. "Agricultural Practice in Texas." *De Bow's Review* 26 (March 1859): 310–312.

Winkler, E. W., ed. "The Bryan-Hayes Correspondence." *Southwestern Historical Quarterly* 25 (October 1921): 98–120.

———, ed. "The Bryan-Hayes Correspondence." *Southwestern Historical Quarterly* 25 (January 1922): 198–221.

———, ed. "The Bryan-Hayes Correspondence," *Southwestern Historical Quarterly* 25 (April 1922): 274–229.

———, ed. "The Bryan-Hayes Correspondence," *Southwestern Historical Quarterly* 26 (July 1922): 184–206.

NEWSPAPERS AND PERIODICALS (FROM TEXAS EXCEPT WHERE INDICATED OTHERWISE)

Anderson Central Texian, 1855–57.

Anderson Texas Baptist, 1855–61.

Bellville Countryman, 1860–64.

Brazoria Advocate of the People's Rights, 1833–34.
Brazoria Brazos Courier, 1839–40.
Brazoria Constitutional Advocate, 1832–33.
Brazoria Texas Planter, 1831, 1852–55.
Brenham Enquirer, 1853–63.
Brooklyn (NY) Daily Eagle.
Columbia Democrat, 1853–55.
Columbia Democrat and Planter, 1855–57.
Columbia Planter, 1843–44.
Columbia Planter's Gazette, 1843.
Galveston News, 1846–61.
Houston Tri-Weekly Telegraph, 1856–64.
New Orleans De Bow's Review.
San Felipe and Houston Telegraph and Texas Register, 1845–53.
Washington American, 1855–57.
Washington Lone Star, 1844–52.
Washington Texas Ranger, 1849–56.

SECONDARY SOURCES

Books

Anzaldúa, Gloria. *Borderlands/Fronteras: The New Mestiza*. 2nd ed. With an introduction by Sonia Saldívar-Hull. 1987. Reprint, San Francisco: Aunt Lute Books, 1999.

Ayers, Edward L. *In the Presence of Mine Enemies: The War in the Heart of America*. New York: Norton, 2003.

———. *Vengeance and Justice: Crime and Punishment in the 19th-Century American South*. New York: Oxford University Press, 1984.

Bade, Klaus J., ed. *Deutsche im Ausland—Fremde in Deutschland: Migration in Geschichte und Gegenwart*. Munich: Verlag C. H. Beck, 1992.

Bailyn, Bernard. *The Ideological Origins of the American Revolution*. Cambridge, MA: Belknap Press of Harvard University Press, 1967.

Bailyn, Bernard, and Philip D. Morgan, eds. *Strangers within the Realm: Cultural Margins of the First British Empire*. Chapel Hill, NC: Institute for Early American History and Culture, 1991.

Baptist, Edward E. *Creating an Old South: Middle Florida's Plantation Frontier before the Civil War*. Chapel Hill: University of North Carolina Press, 2002.

Bardaglio, Peter W. *Reconstructing the Household: Families, Sex, and the Law in the Nineteenth-Century South*. Chapel Hill: University of North Carolina Press, 1995.

Barker, Eugene C. *The Life of Stephen F. Austin, Founder of Texas, 1793–1836*. Austin: University of Texas Press, 1926.

———. *Mexico and Texas, 1821–1835*. New York: Russell, 1965.

Barney, William L. *The Secessionist Impulse: Alabama and Mississippi in 1860*. Princeton, NJ: Princeton University Press, 1974.

Barr, Alwyn. *Black Texans: A History of African Americans in Texas, 1528–1995*. 2nd ed. Norman: University of Oklahoma Press, 1996.

Baum, Dale. *The Shattering of Texas Unionism: Politics in the Lone Star State during the Civil War Era*. Baton Rouge: Louisiana State University Press, 1998.

Bederman, Gail. *Manliness and Civilization: A Cultural History of Gender and Race in the United States, 1880–1917*. Chicago: University of Chicago Press, 1995.

Bennett, Herman L. *Africans in Colonial Mexico: Absolutism, Christianity, and Afro-Creole Consciousness, 1570–1640*. Bloomington: Indiana University Press, 2003.

Bercaw, Nancy D. *Gendered Freedoms: Race, Rights, and the Politics of the Household in the Delta, 1861–1865*. Gainesville: University Press of Florida, 2003.

Bergad, Laird, Fe Iglesias García, and María Del Carmen Barcia. *The Cuban Slave Market, 1790–1880*. Cambridge: Cambridge University Press, 1995.

Berlin, Ira. *Generations of Captivity: A History of African-American Slaves*. Cambridge, MA: Harvard-Belknap, 2003.

———. *Many Thousands Gone: The First Two Centuries of Slavery in North America*. Cambridge, MA: Harvard-Belknap, 1998.

———. *Slaves without Masters: The Free Negro in the Antebellum South*. New York: New Press, 1974.

Berlin, Ira, Barbara J. Fields, Steven F. Miller, Joseph P. Reidy, and Leslie S. Rowland. *Slaves No More: Three Essays on Emancipation and the Civil War*. Cambridge: Cambridge University Press, 1992.

Berlin, Ira, and Philip D. Morgan, eds. *Cultivation and Culture: Labor and the Shaping of Slave Life in the Americas*. Charlottesville: University of Virginia Press, 1992.

———, eds. *The Slaves' Economy: Independent Production by Slaves in the Americas*. London: Frank Cass, 1991.

Binkley, William C. *The Texas Revolution*. Baton Rouge: Louisiana State University Press, 1952; Texas State Historical Association, 1979.

Blackburn, Robin. *The Overthrow of Colonial Slavery, 1776–1848*. London: Verso, 1988.

Bleser, Carol, ed. *In Joy and in Sorrow: Women, Family, and Marriage in the Victorian South, 1830–1900*. New York: Oxford University Press, 1991.

Bodnar, John. *The Transplanted: A History of Immigrants in Urban America*. Bloomington: Indiana University Press, 1985.

Bolton, Charles C. *Poor Whites of the Antebellum South: Tenants and Laborers in Central North Carolina and Northeast Mississippi*. Durham, NC: Duke University Press, 1994.

Boswell, Angela. *Her Act and Deed: Women's Lives in a Rural Southern County, 1837–1873*. College Station: Texas A&M University Press, 2002.

Bowman, Shearer Davis. *Masters and Lords: Mid-Nineteenth Century U.S. Planters and Prussian Junkers*. New York: Oxford University Press, 1993.

Boyer, Richard. *Lives of the Bigamists: Marriage, Family, and Community in Colonial Mexico*. Abridged ed. Albuquerque: University of New Mexico Press, 2000.

Brown, Kathleen M. *Good Wives, Nasty Wenches, and Anxious Patriarchs: Gender, Race, and Power in Colonial Virginia.* Chapel Hill, NC: Institute of Early American History and Culture, 1996.

Buenger, Walter L. *Secession and the Union in Texas.* Austin: University of Texas Press, 1984.

Buenger, Walter L., and Robert A. Calvert. *Texas through Time: Evolving Interpretations.* College Station: Texas A&M University Press, 1991.

Burkhalter, Lois Wood. *Gideon Lincecum, 1793–1874: A Biography.* Austin: University of Texas Press, 1965.

Burton, Orville Vernon. *In My Father's House Are Many Mansions: Family and Community in Edgefield, South Carolina.* Chapel Hill: University of North Carolina Press, 1985.

Camp, Stephanie M. H. *Closer to Freedom: Enslaved Women and Everyday Resistance in the Plantation South.* Chapel Hill: University of North Carolina Press, 2004.

Campbell, Randolph B. *An Empire for Slavery: The Peculiar Institution in Texas, 1821–1865.* Baton Rouge: Louisiana State University Press, 1989.

———. *Gone to Texas: A History of the Lone Star State.* New York: Oxford University Press, 2003.

———. *Grass-Roots Reconstruction in Texas, 1865–1880.* Baton Rouge: Louisiana State University Press, 1997.

———. *Sam Houston and the American Southwest.* New York: Harper Collins, 1993.

———. *A Southern Community in Crisis: Harrison County, Texas, 1850–1880.* Austin: Texas State Historical Association, 1983.

Campbell, Randolph B., and Richard G. Lowe. *Planters and Plain Folk: Agriculture in Antebellum Texas.* Dallas: Southern Methodist University Press, 1987.

———. *Wealth and Power in Antebellum Texas.* College Station: Texas A&M University Press, 1977.

Cantrell, Gregg. *Feeding the Wolf: John B. Rayner and the Politics of Race: 1850–1918.* Wheeling, IL: Harlan Davidson, 2001.

———. *Kenneth and John B. Rayner and the Limits of Southern Dissent.* Urbana: University of Illinois Press, 1993.

———. *Stephen F. Austin: Empresario of Texas.* New Haven, CT: Yale University Press, 1999.

Carroll, Mark M. *Homesteads Ungovernable: Families, Sex, Race, and the Law in Frontier Texas, 1823–1860.* Austin: University of Texas Press, 2001.

Cashin, Joan E. *A Family Venture: Men and Women on the Southern Frontier.* Baltimore: Johns Hopkins University Press, 1991.

Cat Spring Agricultural Society. *The Cat Spring Story.* San Antonio: Lone Star, 1956.

Censer, Jane Turner. *North Carolina Planters and their Children, 1800–1860.* Baton Rouge: Louisiana State University Press, 1984.

———. *The Reconstruction of Southern White Womanhood, 1865–1895.* Baton Rouge: Louisiana State University Press, 2003.

Channing, Steven A. *Crisis of Fear: Secession in South Carolina.* New York: Simon and Schuster, 1970.

Chicoine, Stephen. *The Confederates of Chappell Hill, Texas: Prosperity, Civil War, and Decline.* Jefferson, NC: McFarland, 2005.

Chipman, Donald E. *Spanish Texas, 1519–1821.* Austin: University of Texas Press, 1992.

Clark, Thomas D., and John D. W. Guice. *Frontiers in Conflict: The Old Southwest, 1795–1830.* Albuquerque: University of New Mexico Press, 1989.

Clinton, Catherine. *The Plantation Mistress: Woman's World in the Old South.* New York: Pantheon Books, 1982.

Clinton, Catherine, and Nina Silber, eds., *Divided Houses: Gender and the Civil War.* New York: Oxford University Press, 1992.

Conrad, Robert Edgar. *World of Sorrow: The African Slave Trade to Brazil.* Baton Rouge: Louisiana State University Press, 1986.

Cooper, Frederick, Thomas C. Holt, and Rebecca J. Scott. *Beyond Slavery: Explorations of Race, Labor, and Citizenship in Postemancipation Societies.* Chapel Hill: University of North Carolina Press, 2000.

Cooper, William J., Jr. *Liberty and Slavery: Southern Politics to 1860.* New York: Alfred A. Knopf, 1983.

———. *The South and the Politics of Slavery, 1828–1856.* Baton Rouge: Louisiana State University Press, 1978.

Craton, Michael. *Testing the Chains: The Resistance to Slavery in the British West Indies.* Ithaca. NY: Cornell University Press, 1982.

Creighton, James A. *A Narrative History of Brazoria County.* Waco: Texian Press, 1975.

Cronon, William, George Miles, and Jay Gitlin, eds. *Under an Open Sky: Rethinking America's Western Past.* New York: Norton, 1992.

Crouch, Barry A. *The Freedmen's Bureau and Black Texans.* Austin: University of Texas Press, 1992.

Curtin, Philip D. *The Atlantic Slave Trade: A Census.* Madison: University of Wisconsin Press, 1969.

———. *The Rise and Fall of the Plantation Complex: Essays in Atlantic History.* Cambridge: Cambridge University Press, 1990.

Davis, David Brion. *The Problem of Slavery in the Age of Revolution, 1770–1823.* Ithaca, NY: Cornell University Press, 1975.

Davis, Harvey, ed. *Essays in the Law of Property.* Dallas: Southern Methodist University Press, 1975.

Dean, Warren. *Rio Claro: A Brazilian Plantation System, 1820–1920.* Stanford, CA: Stanford University Press, 1976.

Deyle, Steven. *Carry Me Back: The Domestic Slave Trade in American Life.* New York: Oxford University Press, 2005.

Dietrich, Wilfred O. *The Blazing Story of Washington County.* Brenham, TX: Banner-Press, 1950.

Drescher, Seymour. *Econocide: British Slavery in the Era of Abolition.* Pittsburgh: University of Pittsburgh Press, 1977.

Dubois, Laurent. *Avengers of the New World: The Story of the Haitian Revolution.* Cambridge, MA: Belknap Press of Harvard University, 2004.

———. *A Colony of Citizens: Revolution and Slave Emancipation in the French Caribbean,*

1787–1804. Williamsburg, VA, and Chapel Hill: University of North Carolina Press for the Omohundro Institute of Early American Culture, 2004.

Dupre, Daniel S. *Transforming the Cotton Frontier: Madison County, Alabama, 1800–1840*. Baton Rouge: Louisiana State University Press, 1997.

Edgerton, Douglass. *Gabriel's Rebellion: The Virginia Slave Conspiracies of 1800 and 1802*. Chapel Hill: University of North Carolina Press, 1993.

Edwards, Laura F. *Gendered Strife and Confusion: The Political Culture of Reconstruction*. Urbana: University of Illinois Press, 1997.

Eltis, David. *Economic Growth and the Ending of the Transatlantic Slave Trade*. New York: Oxford University Press, 1987.

———. *The Rise of African Slavery in the Americas*. Cambridge: Cambridge University Press, 2000.

Eltis, David, Stephen D. Behrendt, David Richardson, and Herbert S. Klein, eds. *The Trans-Atlantic Slave Trade: A Database on CD-ROM*. Cambridge: Cambridge University Press, 1999.

Evans, Richard J., and W. R. Lee. *The German Family: Essays on the Social History of the Family in Nineteenth- and Twentieth-Century Germany*. London: Croom Helm, 1981.

Falola, Toyin, and Matt D. Childs, eds. *The Yoruba Diaspora in the Atlantic World*. Bloomington: Indiana University Press, 2004.

Faragher, John Mack, ed. *Rereading Frederick Jackson Turner: "The Significance of the Frontier in American History" and Other Essays*. New York: Henry Holt, 1994.

———. *Sugar Creek: Life on the Illinois Prairie*. New Haven, CT: Yale University Press, 1986.

Faust, Drew Gilpin. *Mothers of Invention: Women of the Slaveholding South in the American Civil War*. New York: Vintage Books, 1996.

———. *Southern Stories: Slaveholders in Peace and War*. Columbia: University of Missouri Press, 1992.

Ferrie, Joseph. *Yankeys Now: Immigrants and the Antebellum United States, 1840–1860*. New York: Oxford University Press, 1999.

Fick, Carolyn E. *The Making of Haiti: The Saint Domingue Revolution from Below*. Knoxville: University of Tennessee Press, 1990.

Fields, Barbara Jeanne. *Slavery and Freedom on the Middle Ground: Maryland in the Civil War*. New Haven, CT: Yale University Press, 1986.

Fish, Stanley. *Is There a Text in This Class? The Authority of Interpretive Communities*. Cambridge, MA: Harvard University Press, 1980.

Fogel, Robert William. *Without Consent or Contract: The Rise and Fall of American Slavery*. New York: Norton, 1989.

Fogel, Robert William, and Stanley L. Engermann. *Time on the Cross: The Economics of American Negro Slavery*. New York: Little, Brown, 1974; Norton, 1989.

Fogel, Robert W., Ralph A. Galantine, and Richard L. Manning. *Without Consent or Contract: The Rise and Fall of American Slavery, Evidence and Methods*. New York: Norton, 1989.

Foley, Neil. *The White Scourge: Mexicans, Blacks, and Poor Whites in Texas Cotton Culture*. Berkeley: University of California Press, 1997.

Follett, Richard. *The Sugar Masters: Planters and Slaves in Louisiana's Cane World, 1820–1860.* Baton Rouge: Louisiana State University Press, 2005.

Foner, Eric. *Free Soil, Free Labor, Free Men: The Ideology of the Republican Party before the Civil War.* New York: Oxford University Press, 1970.

———. *Nothing but Freedom: Emancipation and Its Legacy.* Baton Rouge: Louisiana State University Press, 1983.

———. *Politics and Ideology in the Age of the Civil War.* New York: Oxford University Press, 1980.

———. *Reconstruction: America's Unfinished Revolution, 1863–1877.* New York: Harper and Row, 1988.

———. *The Story of American Freedom.* New York: Norton, 1998.

Ford, Lacy K. *Origins of Southern Radicalism: The South Carolina Upcountry, 1800–1860.* New York: Oxford University Press, 1988.

Fornell, Earl Wesley. *The Galveston Era: The Texas Crescent on the Eve of Secession.* Austin: University of Texas Press, 1961.

Foucault, Michel. *Discipline and Punish: The Birth of the Prison.* Translated by Alan Sheridan. New York: Random House, 1977; Vintage Books, 1979.

———. *The History of Sexuality, an Introduction.* Vol. 1. Translated by Robert Hurley. New York: Random House, 1979; Vintage Books, 1990.

Fox-Genovese, Elizabeth. *Within the Plantation Household: Black and White Women of the Old South.* Chapel Hill: University of North Carolina Press, 1988.

Fox-Genovese, Elizabeth, and Eugene D. Genovese. *Fruits of Merchant Capital: Slavery and Bourgeois Property in the Rise and Expansion of Capitalism.* New York: Oxford University Press, 1983.

Franklin, John Hope, and Loren Schweninger. *Runaway Slaves: Rebels on the Plantation.* New York: Oxford University Press, 1999.

Frederickson, George M. *The Black Image in the White Mind: The Debate on Afro-American Character and Destiny, 1817–1914.* New York: Harper and Row, 1971.

Freehling, William W. *Prelude to Civil War: The Nullification Controversy in South Carolina, 1816–1836.* New York: Harper and Row, 1965.

———. *The Road to Disunion: Secessionists at Bay, 1776–1854.* New York: Oxford University Press, 1990.

———, ed. *Slavery and Freedom.* Expanded ed. New York: Oxford University Press, 1982.

Frey, Sylvia R. *Water from the Rock: Black Resistance in a Revolutionary Age.* Princeton, NJ: Princeton University Press, 1991.

Frey, Sylvia R., and Betty Wood, eds. *From Slavery to Emancipation in the Atlantic World.* London: Frank Cass, 1999.

Friedman, Jean E. *The Enclosed Garden: Women and Community in the Evangelical South, 1830–1900.* Chapel Hill: University of North Carolina Press, 1985.

Frizzell, Isabel. *Bellville: The Founders and Their Legacy.* New Ulm, TX: New Ulm Enterprise, 1992.

Frye-Jacobson, Matthew. *Whiteness of a Different Color: European Immigrants and the Alchemy of Race.* Cambridge, MA: Harvard University Press, 1998.

Galloway, J. H. *The Sugar Cane Industry: An Historical Geography from Its Origins to 1914.* Cambridge: Cambridge University Press 1989.

Gambrell, Herbert. *Anson Jones: The Last President of Texas.* Austin: University of Texas Press, 1947.

Genovese, Eugene D. *From Rebellion to Revolution: Afro-American Slave Revolts in the Making of the Modern World.* Baton Rouge: Louisiana State University Press, 1979.

———. *The Political Economy of Slavery: Studies in the Economy and Society of the Slave South.* New York: Random House, 1961.

———. *Roll, Jordan, Roll: The World the Slaves Made.* New York: Vintage Books, 1972.

———. *The World the Slaveholders Made: Two Essays in Interpretation.* New York: Vintage Books, 1969.

Gish, Theodore, and Richard Spuler. *The Eagle in the New World: German Immigration to Texas and America.* College Station: Texas A&M Press, 1986.

Gomez, Michael A. *Exchanging our Country Marks: The Transformation of African Identities in the Colonial Era and Antebellum South.* Chapel Hill: University of North Carolina Press, 1998.

Goodwyn, Lawrence C. *Democratic Promise: The Populist Moment in America.* New York: Oxford University Press, 1976.

Gray, Lewis Cecil. *History of Agriculture in the Southern United States to 1860.* 2 vols. Washington, DC: Carnegie Institution, 1933.

Griggs, William Clark. *The Elusive Eden: Frank McMullan's Confederate Colony in Brazil.* Austin: University of Texas Press, 1987.

Grimes County Historical Commission. *History of Grimes County: Land of Heritage and Progress.* Navasota, TX: Grimes County Historical Commission, 1982.

Gudmestad, Robert H. *A Troublesome Commerce: The Transformation of the Interstate Slave Trade.* Baton Rouge: Louisiana State University Press, 2003.

Hahn, Steven. *A Nation under Our Feet: Black Political Struggles in the Rural South from Slavery to the Great Migration.* Cambridge: Cambridge University Press, 2003.

———. *The Roots of Southern Populism: Yeoman Farmers and the Transformation of the Georgia Upcountry, 1850–1890.* New York: Oxford University Press, 1983.

Hall, Gwendolyn Midlo. *Africans in Colonial Louisiana: The Development of Afro-Creole Culture in the Eighteenth Century.* Baton Rouge: Louisiana State University Press, 1992.

———. *Slavery and African Ethnicities in the Americas: Restoring the Links.* Chapel Hill: University of North Carolina Press, 2005.

Hall, Jacquelyn Dowd, James Leloudis, Robert Kourstad, Mary Murphy, LuAnn Jones, and Christopher B. Daly. *Like a Family: The Making of a Southern Cotton Mill World.* Chapel Hill: University of North Carolina Press, 1987.

Harris, J. William. *Deep Souths: Delta, Piedmont, and Sea Island Society in the Age of Segregation.* Baltimore: Johns Hopkins University Press, 2001.

———. *Plain Folk and Gentry in a Slave Society: White Liberty and Black Slavery in Augusta's Hinterlands.* Hanover, NH: Wesleyan University Press, 1985.

Hay, Margaret Jean, and Sharon Stichter, eds. *African Women South of the Sahara.* London: Longman, 1984.

Helg, Aline. *Our Rightful Share: The Afro-Cuban Struggle for Equality, 1886–1912.* Chapel Hill: University of North Carolina Press, 1995.

Heyrman, Christine Leigh. *Southern Cross: The Beginnings of the Bible Belt.* New York: Alfred A. Knopf, 1997.

Heywood, Linda M., ed. *Central Africans and Cultural Transformations in the American Diaspora.* Cambridge: Cambridge University Press, 2002.

Hietala, Thomas R. *Manifest Design: Anxious Aggrandizement in Late Jacksonian America.* Ithaca, NY: Cornell University Press, 1985.

Hilton, Anne. *The Kingdom of Kongo.* Oxford: Clarendon Press, 1985.

Himmel, Kelly F. *The Conquest of the Karankawas and the Tonkawas, 1821–1859.* College Station: Texas A&M University Press, 1999.

Hoerder, Dirk, and Jörg Nagler, eds. *People in Transit: German Migrations in Comparative Perspective, 1820–1930.* New York: Cambridge University Press, 1995.

Hogan, William Ransom. *The Texas Republic: A Social and Economic History.* Norman: University of Oklahoma Press, 1946. Reprint, Austin: University of Texas Press, 1969.

Holt, Michael F. *The Political Crisis of the 1850s.* New York: John Wiley, 1978.

Holt, Thomas. *Black over White: Negro Political Leadership in South Carolina during Reconstruction.* Urbana: University of Illinois Press, 1977.

———. *The Problem of Freedom: Race, Labor, and Politics in Jamaica and Britain, 1832–1938.* Baltimore: Johns Hopkins University Press, 1992.

Horsman, Ronald. *Race and Manifest Destiny.* Cambridge, MA: Harvard University Press, 1981.

Hudson, Larry E., Jr. *To Have and to Hold: Slave Work and Family Life in Antebellum South Carolina.* Athens: University of Georgia Press, 1997.

———, ed. *Working toward Freedom: Slave Society and Domestic Economy in the American South.* Rochester, NY: University of Rochester Press, 1994.

Isaac, Rhys. *The Transformation of Virginia, 1740–1790.* Chapel Hill: University of North Carolina Press, 1982; Norton, 1988.

Jaynes, Gerald David. *Branches without Roots: Genesis of the Black Working Class in the American South, 1862–1882.* New York: Oxford University Press, 1986.

Jeffrey, Julie Roy. *Frontier Women: The Trans-Mississippi West, 1840–1880.* New York: Hill and Wang, 1979.

John, Elizabeth H. *Storms Brewed in Other Men's Worlds: The Confrontation of Indians, Spanish, and French in the Southwest, 1540–1795.* 2nd ed. Norman: University of Oklahoma Press, 1996.

Johnson, Michael P. *Toward a Patriarchal Republic: Secession in Georgia.* Baton Rouge: Louisiana State University Press, 1977.

Johnson, Walter. *Soul by Soul: Life inside the Antebellum Slave Market.* Cambridge, MA: Harvard University Press, 1999.

Johnson, William R. *A Short History of the Sugar Industry in Texas.* Houston: Gulf Coast Historical Association, 1961.

Jones, C. Allan. *Texas Roots: Agriculture and Rural Life before the Civil War.* College Station: Texas A&M University Press, 2003.

Jones, Marie Beth. *Peach Point Plantation: The First 150 Years.* Waco, TX: Texian Press, 1982.

Jordan, Terry G. *German Seed in Texas Soil: Immigrant Farmers in Nineteenth-Century Texas.* Austin: University of Texas Press, 1966.

Jordan, Terry G., and Matti Kaups. *The American Backwoods Frontier: An Ethnic and Ecological Interpretation.* Baltimore: Johns Hopkins University Press, 1989.

Jordan, Winthrop D. *Tumult and Silence at Second Creek: An Inquiry into a Civil War Slave Conspiracy.* Rev. ed. Baton Rouge: Louisiana State University Press, 1995.

———. *White over Black: American Attitudes toward the Negro, 1550–1812.* Chapel Hill: University of North Carolina Press, 1968.

Joyner, Charles. *Down by the Riverside: A South Carolina Slave Community.* Urbana: University of Illinois Press, 1984.

Kamphoefner, Walter D. *The Westfalians: From Germany to Missouri.* Princeton, NJ: Princeton University Press, 1987.

Kerber, Linda K. *Women of the Republic: Intellect and Ideology in Revolutionary America.* Chapel Hill: University of North Carolina Press, 1980.

Kerr Ritchie, Jefferey R. *Freedpeople in the Tobacco South: Virginia, 1860–1900.* Chapel Hill: University of North Carolina Press, 1999.

Kierner, Cynthia A. *Beyond the Household: Women's Place in the Early South, 1700–1835.* Ithaca, NY: Cornell University Press, 1998.

Killen, Mrs. James C., ed. *History of Lee County, Texas.* Quanah, TX: Nortex Press, 1974.

Kingston, Mike, Sam Attlesey, and Mary G. Crawford, comps., *The Texas Almanac's Political History of Texas.* Austin: Eakins Press, 1992.

Klein, Herbert, *The Atlantic Slave Trade.* Cambridge: Cambridge University Press, 1999.

Klein, Kerwin Lee. *Frontiers of Historical Imagination: Narrating the European Conquest of Native America, 1890–1990.* Berkeley: University of California Press, 1997.

Knight, Franklin W. *Slave Society in Cuba during the Nineteenth Century.* Madison: University of Wisconsin Press, 1970.

Kolchin, Peter. *American Slavery, 1619–1877.* New York: Hill and Wang, 1993.

———. *First Freedom: The Responses of Alabama's Blacks to Emancipation and Reconstruction.* Westport, CT: Greenwood Press, 1972.

———. *A Sphinx on the American Land: The Nineteenth-Century South in Comparative Perspective.* Baton Rouge: Louisiana State University Press, 2003.

———. *Unfree Labor: American Slavery and Russian Serfdom.* Cambridge, MA: Harvard University Press, 1987.

Kreidte, Peter, Hans Medick, and Jürgen Schlumbohm. *Industrialization before Industrialization: Rural Industry in the Genesis of Capitalism.* Translated by Beate Schempp. With contributions from Herbert Kisch and Franklin F. Mendels. Cambridge: Cambridge University Press, 1981.

Kulikoff, Allan. *Tobacco and Slaves: The Development of Southern Cultures in the Chesapeake, 1680–1800.* Chapel Hill, NC: Institute of Early American History and Culture, 1986.

Lack, Paul D. *The Texas Revolutionary Experience: A Political and Social History, 1835–1836.* College Station: Texas A&M Press, 1992.

Lara, Sylvia Hunold. *Campos da Violencia: Escravos e Senhores na Capitania do Rio de Janeiro, 1750–1808.* Rio de Janeiro: Paz e Terra, 1988.

La Vere, David. *The Texas Indians.* College Station: Texas A&M University Press, 2004.

Law, Robin. *Ouidah: The Social History of a West African Slaving "Port," 1727–1892.* Athens: Ohio University Press, 2004.

———. *The Oyo Empire, c. 1600–1836: A West African Imperialism in the Era of the Atlantic Slave Trade.* Oxford: Clarendon Press, 1977.

———. *The Slave Coast of West Africa, 1550–1750: The Impact of the Atlantic Slave Trade on an African Society.* Oxford: Clarendon, 1991.

Law, Robin, and Silke Stickrodt. *Ports of the Slave Trade (Bights of Benin and Biafra): Papers from a Conference of the Centre of Commonwealth Studies, University of Stirling, June 1998.* Stirling, U.K.: Centre of Commonwealth Studies, 1999.

Lebsock, Suzanne. *The Free Women of Petersburg: Status and Culture in a Southern Town, 1784–1860.* New York: Norton, 1984.

Leubke, Frederick C. *Germans in the New World: Essays in the History of Immigration.* Urbana: University of Illinois Press, 1990.

Limerick, Patricia Nelson. *The Legacy of Conquest: The Unbroken Past of the American West.* New York: Norton, 1987.

Limerick, Patricia Nelson, Clyde A. Milner II, and Charles E. Rankin, eds. *Trails: Toward a New Western History.* Lawrence: University of Kansas Press, 1991.

Lincecum, Jerry Bryan, and Edward Hake Phillips. *Adventures of a Frontier Naturalist: The Life and Times of Dr. Gideon Lincecum.* College Station: Texas A&M University Press, 1994.

Lincecum, Jerry Bryan, Edward Hake Phillips, and Peggy A. Radshaw, eds. *Gideon Lincecum's Sword: Civil War Letters from the Texas Home Front.* Denton: University of North Texas Press, 2001.

Lindemann, Ann, James Lindemann, and William Richter. *Historical Accounts of Industry: Texas, 1831–1986.* New Ulm, TX: New Ulm Enterprise, 1986.

Loveland, Ann C. *Southern Evangelicals and the Social Order, 1800–1860.* Baton Rouge: Louisiana State University Press, 1980.

Lowe, Richard G., and Randolph B. Campbell. *Planters and Plain Folk: Agriculture in Antebellum Texas.* Dallas: Southern Methodist University Press, 1987.

Malone, Ann Patton. *Sweet Chariot: Slave Family and Household Structure in Nineteenth-Century Louisiana.* Chapel Hill: University of North Carolina Press, 1992.

Manning, Patrick. *Slavery, Colonialism, and Economic Growth in Dahomey, 1640–1960.* Cambridge: Cambridge University Press, 1982.

Marten, James. *Texas Divided: Loyalty and Dissent in the Lone Star State, 1856–1874.* Lexington: University Press of Kentucky, 1990.

Martin, Jonathan. *Divided Mastery: Slave Hiring in the Antebellum South.* Cambridge, MA: Harvard University Press, 2004.

Martin, Phyllis. *The External Trade of the Loango Coast, 1576–1870.* Oxford: Clarendon Press, 1972.

Martin, Roscoe. *The People's Party in Texas: A Study in Third Party Politics.* Austin: University of Texas Press, 1933.

Mathews, Donald G. *Religion in the Old South.* Chicago: University of Chicago Press, 1977.

McCardell, John. *The Idea of a Southern Nation: Southern Nationalists and Southern Nationalism, 1830–1860.* New York: Norton, 1979.

McCormick, Andrew Phelps. *The Scotch-Irish in America, as Shown in Sketches of the Pioneer Scotch-Irish Families McCormick, Stevenson, McKenzie, and Bell, in North Carolina, Kentucky, Missouri, and Texas.* New Orleans: N.p., 1897.

McCurry, Stephanie. *Masters of Small Worlds: Yeoman Households, Gender Relations, and the Political Culture of the Antebellum South Carolina Low Country.* New York: Oxford University Press, 1995.

McDonald, Roderick A. *The Economy and Material Culture of Slaves: Goods and Chattels on the Sugar Plantations of Jamaica and Louisiana.* Baton Rouge: Louisiana State University Press, 1993.

Meinig, Donald W. *Imperial Texas: An Interpretive Essay in Cultural Geography.* Austin: University of Texas Press, 1969.

Melish, Joanne Pope. *Disowning Slavery: Gradual Emancipation and "Race" in New England, 1780–1860.* Ithaca, NY: Cornell University Press, 1998.

Michaelson, Scott, and David E. Johnson, eds. *Border Theory: The Limits of Cultural Politics.* Minneapolis: University of Minnesota Press, 1997.

Miller, Joseph. *Way of Death: Merchant Capitalism and the Angolan Slave Trade, 1730–1830.* Madison: University of Wisconsin Press, 1988.

Miller, Randall, ed. *States of Progress: Germans and Blacks over Three Hundred Years. Lectures from the Tricentennial of the German Protest against Slavery.* Philadelphia: German Society of Pennsylvania, 1989.

Miller, Thomas Lloyd. *The Public Lands of Texas, 1519–1970.* Norman: University of Oklahoma Press, 1972.

Mintz, Sidney W. *Sweetness and Power: The Place of Sugar in Modern History.* New York: Penguin Books, 1985.

Mintz, Sidney W., and Richard Price. *The Birth of African-American Culture: An Anthropological Perspective.* Boston: Beacon Press, 1992. Originally published as *An Anthropological Approach to the Afro-American Past* (Philadelphia: Institute for the Study of Human Issues, 1976).

Mintz, Sidney W., and Sally Price, eds. *Caribbean Contours.* Baltimore: Johns Hopkins University Press, 1985.

Mintz, Steven, and Susan Kellogg. *Domestic Revolutions: A Social History of American Family Life.* New York: Free Press, 1988.

Moneyhon, Carl H. *Republicanism in Reconstruction Texas.* Austin: University of Texas Press, 1980.

———. *Texas after the Civil War: The Struggle for Reconstruction.* College Station: Texas A&M University Press, 2004.

Montejano, David. *Anglos and Mexicans in the Making of Texas, 1836–1986.* Austin: University of Texas Press, 1987.

Moore, John Hebron. *The Emergence of the Cotton Kingdom in the Old Southwest: Mississippi, 1770–1860.* Baton Rouge: Louisiana State University Press, 1988.

Moreno Fraginals, Manuel. *El Ingenio: Complejo Económico Social Cubano del Azucár.* 3 vols. Havana, 1978.

Morgan, Edmund S. *American Slavery, American Freedom: The Ordeal of Colonial Virginia.* New York: Norton, 1975.

Morgan, Philip D. *Slave Counterpoint: Black Culture in the Eighteenth-Century Chesapeake and Lowcountry.* Chapel Hill, NC: Omohundro Institute of Early American History and Culture, 1998.

Morley, David, and Kuan-Hsing Chen. *Stuart Hall: Critical Dialogues in Cultural Studies.* London: Routledge, 1996.

Morris, Christopher C. *Becoming Southern: The Evolution of a Way of Life, Warren County and Vicksburg, Mississippi, 1770–1860.* New York: Oxford University Press, 1995.

Morris, Thomas D. *Southern Slavery and the Law, 1619–1860.* Chapel Hill: University of North Carolina Press, 1996.

Mullin, Gerald W. *Flight and Rebellion: Slave Resistance in Eighteenth-Century Virginia.* New York: Oxford University Press, 1972.

Mullin, Michael. *Africa in America: Slave Acculturation and Resistance in the American South and British Caribbean, 1736–1831.* Urbana: University of Illinois Press, 1991.

Newcomb, W. W., Jr. *The Indians of Texas: From Prehistoric to Modern Times.* Austin: University of Texas Press, 1961.

Oakes, James. *The Ruling Race: A History of American Slaveholders.* New York: Vintage Books, 1982.

———. *Slavery and Freedom: An Interpretation of the Old South.* New York: Vintage Books, 1990.

Olwell, Robert. *Masters, Slaves, and Subjects: The Culture of Power in the South Carolina Low Country, 1740–1790.* Ithaca, NY: Cornell University Press, 1998.

Owsley, Frank Lawrence. *King Cotton Diplomacy: Foreign Relations of the Confederate States of America.* 2nd ed., revised by Harriet Owsley. Chicago: University of Chicago Press, 1959.

———. *Plain Folk of the Old South.* Baton Rouge: Louisiana State University Press, 1949.

Owsley, Frank Lawrence, Jr., and Gene A. Smith. *Filibusters and Expansionists: Jeffersonian Manifest Destiny, 1800–1821.* Tuscaloosa: University of Alabama Press, 1998.

Oyewumi, Oyeronke. *The Invention of Women: Making an African Sense of Western Gender Discourses.* Minneapolis: University of Minnesota Press, 1997.

Palmer, Colin A. *Slaves of the White God: Blacks in Mexico, 1570–1650.* Cambridge, MA: Harvard University Press, 1976.

Penningroth, Dylan C. *The Claims of Kinfolk: African American Property and Community in the Nineteenth-Century South.* Chapel Hill: University of North Carolina Press, 2003.

Pennington, Mrs. R. E. *The History of Brenham and Washington County.* Houston: Standard, 1915.

Phillips, Ulrich B. *Plantation and Frontier, 1649–1863.* 2 vols. Cleveland: Arthur H. Clark, 1910.

Potter, David M., *The Impending Crisis, 1848–1861.* Completed and edited by Don E. Fehrenbacher. New York: Harper and Row, 1976.

Price, Richard, ed. *Maroon Societies: Rebel Slave Communities in the Americas.* New York: Anchor Books, 1973; Johns Hopkins University Press, 1979.

Puryear, Pamela, and Nath Winfield Jr. *Sandbars and Sternwheelers: Steam Navigation on the Brazos.* College Station: Texas A&M University Press, 1976.

Randers-Pehrson, Justine Davis. *Adolf Douai, 1819–1888: The Turbulent Life of a German Forty-Eighter in the Homeland and in the United States.* New York: Peter Lang, 2000.

Ransom, Roger L., and Richard Sutch. *One Kind of Freedom: The Economic Consequences of Emancipation.* New York: Cambridge University Press, 1977.

Red, William Stuart. *The Texas Colonists and Religion, 1821–1836.* Austin: Shettles, 1924.

Reed, S. G. *A History of Texas Railroads.* 1941. Reprint, New York: Arno Press, 1981.

Reidy, Joseph P. *From Slavery to Agrarian Capitalism in the Cotton Plantation South: Central Georgia, 1800–1880.* Chapel Hill: University of North Carolina Press, 1992.

Reis, João José. *Slave Rebellion in Brazil: The Muslim Uprising of 1835 in Bahia.* Baltimore: Johns Hopkins University Press, 1993.

Reséndez, Andrés. *Changing National Identities at the Frontier: Texas and New Mexico, 1800–1850.* Cambridge: Cambridge University Press, 2005.

Rice, Lawrence. *The Negro in Texas, 1874–1900.* Baton Rouge: Louisiana State University Press, 1971.

Richter, William L. *The Army in Texas during Reconstruction, 1865–1870.* College Station: Texas A&M University Press, 1987.

———. *Overreached on all Sides: The Freedmen's Bureau Administrators in Texas, 1865–1868.* College Station: Texas A&M University Press, 1991.

Rickliss, Robert. *The Karankawa Indians of Texas: An Ecological Study of Cultural Tradition and Change.* Austin: University of Texas Press, 1996.

Rivers, Larry Eugene. *Slavery in Florida: Territorial Days to Emancipation.* Gainesville: University Press of Florida, 2000.

Roark, James L. *Masters without Slaves: Southern Planters in the Civil War and Reconstruction.* New York: Norton, 1977.

Roediger, David R. *Towards the Abolition of Whiteness: Essays on Race, Politics, and Working Class History.* New York: Verso, 1994.

———. *The Wages of Whiteness: Race and the Making of the American Working Class.* New York: Verso, 1991.

Rothman, Adam. *Slave Country: American Expansion and the Origins of the Deep South.* Cambridge, MA: Harvard University Press, 2005.

Rubin, Vera, and Arthur Tuden, eds. *Comparative Perspectives on Slavery in New World Plantation Societies.* New York: New York Academy of Sciences, 1977.

Sabean, David Warren. *Power in the Blood: Popular Culture and Village Discourse in Early Modern Germany.* Cambridge: Cambridge University Press, 1984.

———. *Property, Production, and Family in Neckarhausen, 1700–1870.* Cambridge: Cambridge University Press, 1990.

Said, Edward W. *Orientalism.* New York: Random House, 1978; Vintage Books, 1979.

Saville, Julie. *The Work of Reconstruction: From Slave to Wage Laborer in South Carolina, 1860–1870.* New York: Cambridge University Press, 1994.

Saxton, Alexander. *The Rise and Fall of the White Republic: Class Politics and Mass Culture in Nineteenth-Century America.* London: Verso, 1990.

Schlumbohm, Jürgen. *Lebensläufe, Familien, Höfe: Die Bauern und Heuerleute des Osnabrückischen Kirschspiels Belm in proto-industrieller Zeit, 1650–1860*. Göttingen: Vandenhoeck und Ruprecht, 1994.

Schmidt, Charles Frank. *History of Washington County*. San Antonio: Naylor, 1949.

Schuchalter, Jerry. *Frontier and Utopia in the Fiction of Charles Sealsfield: A Study of the Lebensbilder aus der westlichen Hemisphäre*. Frankfurt: Peter Lang, 1986.

Schwalm, Leslie A. *A Hard Fight for We: Women's Transition from Slavery to Freedom in South Carolina*. Urbana: University of Illinois Press, 1997.

Schwartz, Marie Jenkins. *Born in Bondage: Growing Up Enslaved in the Antebellum South*. Cambridge, MA: Harvard University Press, 2000.

Schwartz, Stuart B. *Sugar Plantations in the Formation of Brazilian Society: Bahia, 1550–1835*. Cambridge: Cambridge University Press, 1985.

Scott, Joan Wallach. *Gender and the Politics of History*. New York: Columbia University Press, 1988.

Scott, Rebecca J. *Degrees of Freedom: Louisiana and Cuba after Slavery*. Cambridge, MA: Belknap Press of Harvard University Press, 2005.

———. *Slave Emancipation in Cuba: The Transition to Free Labor, 1865–1899*. Princeton, NJ: Princeton University Press, 1986.

Sellers, Charles. *The Market Revolution: Jacksonian America, 1815–1846*. New York: Oxford University Press, 1991.

Sensbach, Jon F. *A Separate Canaan: The Making of an Afro-Moravian World in North Carolina, 1763–1840*. Chapel Hill: University of North Carolina Press, 1998.

Sidbury, James. *Plowshares into Swords: Race, Rebellion, and Identity in Gabriel's Virginia, 1730–1810*. Cambridge: Cambridge University Press, 1997.

Siegel, Stanley. *A Political History of the Texas Republic, 1836–1845*. Austin: University of Texas Press, 1956.

Silbey, Joel H. *Storm over Texas: The Annexation Controversy and the Road to Civil War*. New York: Oxford University Press, 2005.

Sitterson, J. Carlyle. *Sugar Country: The Cane Sugar Industry in the South, 1753–1950*. Lexington: University of Kentucky Press, 1953.

Sitton, Thad, and Dan K. Utley. *From Can See to Can't: Texas Cotton Farmers on the Southern Prairies*. Austin: University of Texas Press, 1997.

Slotkin, Richard. *The Fatal Environment: The Myth of the Frontier in the Age of Industrialization, 1800–1890*. New York: Atheneum, 1985; Harper Perennial, 1994.

Smallwood, James. *Time of Hope, Time of Despair: Black Texans during Reconstruction*. Port Washington, NY: National University Publications, 1981.

Smallwood, James, Barry A. Crouch, and Larry Peacock. *Murder and Mayhem: The War of Reconstruction in Texas*. College Station: Texas A&M University Press, 2003.

Smith, Mark M. *Mastered by the Clock: Time, Slavery, and Freedom in the American South*. Chapel Hill: University of North Carolina Press, 1997.

Sollors, Werner. *Beyond Ethnicity: Consent and Descent in American Culture*. New York: Oxford University Press, 1986.

Sowell, A. J. *The History of Fort Bend County.* Houston: W. H. Coyle, 1904.

Spratt, John Stricklin. *The Road to Spindletop: Economic Change in Texas, 1875–1901.* Austin: University of Texas Press, 1955.

Steady, Filomena Chioma, ed. *The Black Woman Cross-Culturally.* Cambridge, MA: Schwenkman, 1981.

Stein, Stanley J. *Vassouras, a Brazilian Coffee County: The Roles of Master and Slave in a Plantation Society.* Princeton, NJ: Princeton University Press, 1957.

Stern, Steve J. *The Secret History of Gender: Women, Men, and Power in Late Colonial Mexico.* Chapel Hill: University of North Carolina Press, 1995.

Stevenson, Brenda E. *Life in Black and White: Family and Community in the Slave South.* New York: Oxford University Press, 1996.

Stone, Lawrence. *The Family, Sex, and Marriage in England, 1500–1800.* New York: Harper and Row, 1977.

Stowe, Steven M. *Intimacy and Power in the Old South: Ritual in the Lives of the Planters.* Baltimore: Johns Hopkins University Press, 1987.

Strobel, Abner J. *The Old Plantations and their Owners.* Houston: Union National Bank, 1930.

Struve, Walter. *Germans and Texans: Commerce, Migration, and Culture in the Days of the Lone Star Republic.* Austin: University of Texas Press, 1996.

Stuckey, Sterling. *Slave Culture: Nationalist Theory and the Foundations of Black America.* New York: Oxford University Press, 1987.

Tadman, Michael. *Speculators and Slaves: Masters, Traders, and Slaves in the Old South.* Madison: University of Wisconsin Press, 1989.

Thomas, Hugh. *The Slave Trade: The Story of the Atlantic Slave Trade, 1440–1870.* New York: Simon and Schuster, 1997.

Thompson, E. P. *The Making of the English Working Class.* New York: Random House, 1963; Vintage Books, 1966.

Thompson, Robert Farris. *Flash of the Spirit: African and Afro-American Art and Philosophy.* New York: Random House, 1983; Vintage Books, 1984.

Thornton, J. Mills, III. *Politics and Power in a Slave Society: Alabama, 1800–1860.* Baton Rouge: Louisiana State University Press, 1978.

Thornton, John. *Africa and Africans in the Making of the Atlantic World, 1400–1680.* Cambridge: Cambridge University Press, 1992.

———. *The Kingdom of Kongo: Civil War and Transition, 1641–1718.* Madison: University of Wisconsin Press, 1983.

Trelease, Allen W. *White Terror: The Ku Klux Klan Conspiracy and Southern Reconstruction.* New York: Harper and Row, 1971.

Turner, Mary S. *Slaves and Missionaries: The Disintegration of Jamaican Slave Society, 1787–1834.* Urbana: University of Illinois Press, 1982.

Usner, Daniel H., Jr. *Indians, Settlers, and Slaves in a Frontier Exchange Economy: The Lower Mississippi Valley before 1783.* Chapel Hill: Omohundro Institute of Early American History and Culture, 1992.

Vernon, Walter N., Robert W. Sledge, Robert C. Monk, and Norman W. Spellman. *The Methodist Excitement in Texas: A History.* Dallas: Texas United Methodist Historical Society, 1984.

Walker, Donald R. *Penology for Profit: A History of the Texas Prison System, 1867–1912.* College Station: Texas A&M University Press, 1988.

Walker, Mack. *German Home Towns: Community, State, and General Estate, 1648–1871.* Ithaca, NY: Cornell University Press, 1971.

———. *Germany and the Emigration, 1816–1885.* Cambridge, MA: Harvard University Press, 1964.

Wallerstein, Immanuel. *Historical Capitalism.* London: Verso, 1983.

———. *The Modern World-System: Capitalist Agriculture and the Origins of the European World-Economy in the Sixteenth Century.* New York: Academic Press, 1974.

Walsh, Lorena. *From Calabar to Carter's Grove: The History of a Virginia Slave Community.* Charlottesville: University of Virginia Press, 1997.

Wayne, Michael. *The Reshaping of Plantation Society: The Natchez District, 1860–1880.* Urbana: University of Illinois Press, 1983.

Weber, David J. *The Mexican Frontier, 1821–1846: The American Southwest under Mexico.* Albuquerque: University of New Mexico Press, 1982.

———. *The Spanish Frontier in North America.* New Haven, CT: Yale University Press, 1992.

Weiner, Marli F. *Mistresses and Slaves: Plantation Women in South Carolina, 1830–1880.* Urbana: University of Illinois Press, 1998.

Wharton, Clarence R. *History of Fort Bend County.* Houston: Anson Jones Press, 1939.

White, Deborah Gray. *Ar'n't I a Woman? Female Slaves in the Plantation South.* New York: Norton, 1985.

White, Richard. *The Middle Ground: Indians, Empires, and Republics in the Great Lakes Region, 1650–1815.* Cambridge: Cambridge University Press, 1991.

Williams, Eric. *Capitalism and Slavery.* Chapel Hill: University of North Carolina Press, 1944.

Williamson, Joel. *The Crucible of Race: Black/White Relations in the American South since Emancipation.* New York: Oxford University Press, 1984.

Winfield, Mr. and Mrs. Nath. *All Our Yesterdays: A Brief History of Chappell Hill.* Waco, Tex.: Texian Press, 1969.

Wolf, Eric R. *Europe and the People without History.* Berkeley: University of California Press, 1982.

Wood, Betty. *Women's Work, Men's Work: The Informal Economies of Lowcountry Georgia.* Athens: University of Georgia Press, 1995.

Wood, Gordon S. *The Creation of the American Republic, 1776–1787.* New York: Norton, 1969.

———. *The Radicalism of the American Revolution.* New York: Alfred A. Knopf, 1991.

Wood, Peter. *Black Majority: Negroes in Colonial South Carolina from 1670 to the Stono Rebellion.* New York: Alfred A. Knopf, 1974.

Woodman, Harold D. *King Cotton and His Retainers: Financing and Marketing the Cotton Crop of the South, 1800–1925.* Lexington: University of Kentucky Press, 1968.

———. *New South—New Law: The Legal Foundations of Credit and Labor Relations in the Postbellum Agricultural South.* Baton Rouge: Louisiana State University Press, 1995.

Wright, Gavin. *Old South, New South: Revolutions in the Southern Economy since the Civil War.* New York: Basic Books, 1986.

———. *The Political Economy of the Cotton South: Households, Markets, and Wealth in the Nineteenth Century.* New York: Norton, 1978.

Wunder, Heide. *He Is the Sun, She Is the Moon: Women in Early Modern Germany.* Translated by Thomas Dunlap. Cambridge, MA: Harvard University Press, 1998.

Wyatt-Brown, Bertram. *Southern Honor: Ethics and Behavior in the Old South.* New York: Oxford University Press, 1982.

Yans-McLaughlin, Virginia, ed. *Immigration Reconsidered: History, Sociology, and Politics.* New York: Oxford University Press, 1990.

York, Miriam Korff. *Friedrich Ernst of Industry: Research on Life, Family, Acquaintances, and Conditions of the Times.* N.p., 1989.

Young, Jeffrey Robert. *Domesticating Slavery: The Master Class in Georgia and South Carolina, 1670–1837.* Chapel Hill: University of North Carolina Press, 1999.

Articles and Essays

Addington, Wendell G. "Slave Insurrections in Texas." *Journal of Negro History* 35 (October 1950): 408–434.

Adelman, Jeremy, and Stephen Aron. "From Borderlands to Borders: Empires, Nation-States, and the Peoples In Between in North American History." *American Historical Review* 104 (June 1999): 814–841.

Anderson, Timothy G. "Proto-Industrialization, Sharecropping, and Outmigration in Nineteenth-Century Rural Westphalia." *Journal of Peasant Studies* 29 (October 2001): 1–30.

Ashford, Gerald. "Jacksonian Liberalism and Spanish Law in Early Texas." *Southwestern Historical Quarterly* 57 (July 1953): 1–37.

Barker, Eugene C. "The African Slave Trade in Texas." *Quarterly of the Texas State Historical Association* 6 (October 1902): 145–158.

Barr, Alwyn. "Reconstruction Change and Continuity in Brazoria County, Texas." *Houston Review* 18 (Spring 1996):114–123.

———. "The Texas Black Uprising Scare of 1883." *Phylon* 41 (June 1980): 179–186.

Baughman, James P. "The Evolution of Rail-Water Systems of Transportation in the Gulf Southwest, 1836–1890." *Journal of Southern History* 34 (August 1968): 357–381.

Berlin, Ira. "Time, Space, and the Evolution of Afro-American Society on British Mainland North America." *American Historical Review* 85 (February 1980): 44–78.

Blum, Elizabeth D. "Power, Danger, and Control: Slave Women's Perceptions of Wilderness in the Nineteenth Century." *Women's Studies* 31 (January–February 2002): 247–266.

Borscheid, Peter. "Romantic Love or Material Interest: Choosing Partners in Nineteenth-Century Germany." *Journal of Family History* 11 (April 1986): 157–168.

Brown, Kenneth L. "Ethnographic Analogy, Archaeology, and the African Diaspora: Perspectives from a Tenant Community." *Historical Archaeology,* 38, no. 1 (2004): 79–89.

Brown, Kenneth L., and Doreen C. Cooper. "Structural Continuity in an African-American Slave and Tenant Community." *Historical Archaeology* 24, no. 4 (1990): 7–19.

Buenger, Walter L. "Texas and the Riddle of Secession." *Southwestern Historical Quarterly* 87 (October 1983): 151–182.

Bugbee, Lester G. "The Old Three Hundred: A List of Settlers in Austin's First Colony." *Quarterly of the Texas State Historical Association* 1 (October 1897): 108–120.

Campbell, Randolph. "Intermittent Slave Ownership: Texas as a Test Case." *Journal of Southern History* 60 (February 1985): 15–23.

———. "Political Conflict within the Southern Consensus: Harrison County, Texas, 1850–1880." *Civil War History* 26 (September 1980): 218–239.

———. "Slave Hiring in Texas." *American Historical Review* 93 (February 1988): 107–114.

———. "The Whig Party of Texas in the Elections of 1848 and 1852." *Southwestern Historical Quarterly* 73 (July 1969): 17–34.

Cantrell, Gregg. "Sam Houston and the Know-Nothings: A Reappraisal." *Southwestern Historical Quarterly* 96 (January 1993): 326–343.

Carrigan, William Dean. "Slavery on the Frontier: The Peculiar Institution in Central Texas." *Slavery and Abolition* 20 (August 1999): 63–86.

Conzen, Kathleen Neils, David A. Gerber, Ewa Morawska, George E. Pozzetta, and Rudolph J. Vecoli. "The Invention of Ethnicity: A Perspective from the U.S.A." *Journal of American Ethnic History* 12 (Fall 1992): 3–41.

Crouch, Barry A. "The 'Chords of Love': Legalizing Black Marital and Family Rights in Postwar Texas." *Journal of Negro History* 79 (September 1994): 334–351.

———. "Crisis in Color: Racial Separation in Texas during Reconstruction." *Civil War History* 16 (March 1970): 37–49.

———. "Guardian of the Freedpeople: Texas Freedmen's Bureau Agents and the Black Community." *Southern Studies* 3 (Fall 1992): 185–201.

———. "Hesitant Recognition: Texas Black Politicians, 1865–1900." *East Texas Historical Journal* 31 (Spring 1993): 41–58.

———. "Self Determination and Local Black Leaders in Texas." *Phylon* 39 (December 1978): 344–355.

———. "A Spirit of Lawlessness: White Violence, Texas Blacks, 1865–1868." *Journal of Social History* 18 (Winter 1984): 217–232.

Crouch, Barry A., and Larry Madaras. "Reconstructing Black Families: Perspectives from the Texas Freedmen's Bureau Records." *Prologue* 18 (Summer 1986): 109–122.

Curlee, Abigail. "The History of a Texas Slave Plantation, 1831–63." In *Plantation, Town, and County: Essays on the Local History of American Slave Society*, edited by Elinor Miller and Eugene D. Genovese, 303–334. Urbana: University of Illinois Press, 1974.

Davis, David Brion. "Looking at Slavery from Broader Perspectives." *American Historical Review* 105 (April 2000): 452–466.

Ellenberger, Matthew. "Illuminating the Lesser Lights: Notes on the Life of Albert Clinton Horton." *Southwestern Historical Quarterly* 88 (April 1985): 363–386.

Ellis, L. Tuffly. "The Revolutionizing of the Texas Cotton Trade, 1865–1885." *Southwestern Historical Quarterly* 63 (April 1970): 478–507.

Eltis, David. "The Diaspora of Yoruba Speakers, 1650–1850." In Falola and Childs, *The Yoruba Diaspora in the Atlantic World*, 17–39.

———. "Labour and Coercion in the English Atlantic World from the Seventeenth to the

Twentieth Century." *Slavery and Abolition* 14 (April 1993): 207–226.

———. "The U.S. Transatlantic Slave Trade, 1644–1867: An Assessment." *Civil War History* 54 (December 2008): 347–378.

Eltis, David, and David Richardson. "The 'Numbers Game' and Routes to Slavery." *Slavery and Abolition* 18 (April 1997): 3–15.

———. "West Africa and the Transatlantic Slave Trade: New Evidence of Long-Run Trends." *Slavery and Abolition* 18 (April 1997): 16–35.

Engerman, Stanley L. "Slavery at Different Times and Places." *American Historical Review* 105 (April 2000): 480–484.

Ford, Lacy K., Jr. "Inventing the Concurrent Majority: Madison, Calhoun, and the Problem of Majoritarianism in American Political Thought." *Journal of Southern History* 60 (February 1994): 19–58.

Formisano, Ronald P. "Deferential-Participant Politics: The Early Republic's Political Culture, 1789–1840." *American Political Science Review* 68 (June 1974): 473–487.

Genovese, Eugene D. "Yeomen Farmers in a Slaveholders' Democracy." *Agricultural History* 49 (April 1975): 331–342.

Goodwyn, Lawrence. "Populist Dreams and Negro Rights: East Texas as a Case Study." *American Historical Review* 76 (December 1971): 1135–1171.

Groce, Rosa Bertleth. "Jared Ellison Groce." *Southwestern Historical Quarterly* 20 (April 1917): 358–368.

Haefeli, Evan. "A Note on the Use of North American Borderlands." *American Historical Review* 104 (October 1999): 1222–1225.

Hilliard, Sam B. "Site Characteristics and Spatial Stability of the Louisiana Sugarcane Industry." *Agricultural History* 53 (January 1979): 254–269.

Holbrook, Abigail Curlee. "Cotton Marketing in Antebellum Texas." *Southwestern Historical Quarterly* 63 (April 1970): 431–455.

———. "A Glimpse of Life on Antebellum Slave Plantations in Texas." *Southwestern Historical Quarterly* 76 (April 1973): 361–383.

Howard, Vicki. "The Courtship Letters of an African American Couple: Race, Gender, Class, and the Cult of True Womanhood." *Southwestern Historical Quarterly* 100 (July 1996): 65–80.

Jordan, Terry G. "Germans and Blacks in Texas." In Miller, *States of Progress*, 89–97.

———. "The German Settlement of Texas after 1865." *Southwestern Historical Quarterly* 73 (October 1969): 193–212.

Jordan-Bychkov, Terry, Allen R. Branum, and Paula K. Hood, eds. "The Boesel Letters: Two Texas Germans in Sibley's Brigade." Translated by Irma Ohlendorf Scwarz. *Southwestern Historical Quarterly* 102 (April 1999): 457–484.

Juricek, John T. "American Usage of the Word 'Frontier' from Colonial Times to Frederick Jackson Turner." *Proceedings of the American Philosophical Society* 110 (1966): 10–34.

Kamphoefner, Walter D. "German-Americans and Civil War Politics: A Reconsideration of the Ethnocultural Thesis." *Civil War History* 37 (1991): 232–246.

———. "German Emigration Research, North, South, and East: Findings, Methods, and Open Questions." In Hoerder and Nagler, *People in Transit*, 14–34.

———. "German Texans: In the Mainstream or Backwaters of Lone Star Society?" *Yearbook of German-American Studies* 38 (2003): 119–138.

———. "New Perspectives on Texas Germans and the Confederacy." *Southwestern Historical Quarterly* 102 (April 1999): 441–455.

Kattner, Lauren Ann. "The Diversity of Old South White Women: The Peculiar Worlds of German American Women." In *Discovering the Women in Slavery: Emancipating Perspectives on the American Past,* edited by Patricia Morton, 299–311. Athens: University of Georgia Press, 1996.

Kelley, Sean M. "Blackbirders and *Bozales:* African-Born Slaves on the Lower Brazos River of Texas in the Nineteenth Century." *Civil War History* 54 (December 2008): 406–423.

———. "'Mexico in His Head': Slavery and the Texas-Mexico Border, 1810–1860." *Journal of Social History* 37 (Spring 2004): 708–723.

———. "A Texas Peasantry? Black Smallholders in the Texas Sugar Bowl, 1865–1890." *Slavery and Abolition* 28 (August 2007): 193–209.

Klein, Herbert S. "The Structure of the Atlantic Slave Trade in the Nineteenth Century: An Assessment." *Outre-Mers: Revue d'Histoire* 89 (2002): 63–77.

Koch, Ranier. "Liberalismus, Konservatismus, und das Problem der Negersklaverei: Ein Beitrag zur Geschichte der politischen Denkens in Deutschland in der ersten Hälfte des 19. Jahrhunderts." *Historische Zeitschrift* 222 (June 1976): 529–577.

Kolchin, Peter. "The Big Picture: A Comment on David Brion Davis's 'Looking at Slavery from other Perspectives.'" *American Historical Review* 105 (April 2000): 467–471.

Köllmann, Wolfgang, and Peter Marschalck. "German Emigration to the United States." Translated by Thomas C. Childers. *Perspectives in American History* 7 (1973): 499–554.

Lack, Paul D. "Slavery and the Texas Revolution." *Southwestern Historical Quarterly* 89 (October 1985): 181–202.

Laslett, Peter. "The Character of Familial History, Its Limitations, and the Conditions for Its Proper Pursuit." *Journal of Family History* 12 (January 1987): 263–284.

Lathrop, Barnes F. "Migration into East Texas, 1835–1860." *Southwestern Historical Quarterly* 52 (July 1948): 1–29; 52 (October 1948): 184–208; 52 (January 1949): 325–348.

La Vere, David. "Between Kinship and Capitalism: French and Spanish Rivalry in the Louisiana-Texas Indian Trade." *Journal of Southern History* 64 (May 1998): 197–218.

Ledbetter, Nan Thompson. "The Muddy Brazos in Early Texas." *Southwestern Historical Quarterly* 63 (October 1959): 238–262.

Lovett, Leslie A. "Biracial Politics and Community Development: The Reconstruction Experience in Fort Bend County, Texas, 1869–1889." *Houston Review* 16 (Fall 1994): 27–39.

Malone, Ann Patton. "Matt Gaines: Reconstruction Politician." In *Black Leaders: Texans for Their Times,* edited by Alwyn Barr and Robert A. Calvert, 49–81. Austin: Texas State Historical Association, 1981.

McDaniel, Antonio. "The Power of Culture: A Review of the Idea of Africa's Influence on Family Structure in Antebellum America." *Journal of Family History* 15 (April 1990): 225–238.

McKnight, Joseph W. "Spanish Law for the Protection of Surviving Spouses." *Publicaciones del Instituto Nacional de Estudios Jurâidicos,* ser. 1, no. 4 (1987): 365–406.

———. "Texas Community Property Law—Its Course of Development and Reform." In Davis, *Essays in the Law of Property*, 30–57.

Morgan, Philip D. "The Cultural Implications of the Atlantic Slave Trade: African Regional Origins, American Destinations, and New World Developments." *Slavery and Abolition* 18 (April 1997): 122–145.

Narrett, David E. "José Bernardo Gutiérrez de Lara: *Caudillo* of the Mexican Republic in Texas." *Southwestern Historical Quarterly* 106 (October 2002): 195–228.

Nash, A. E. Keir. "The Texas Supreme Court and Trial Rights of Blacks, 1845–1860." *Journal of American History* 58 (December 1971): 622–642.

Nieman, Donald G. "African American Communities, Politics, and Justice: Washington County, Texas, 1865–1890." In *Local Matters: Race, Crime, and Justice in the Nineteenth-Century South*, edited by Christopher Waldrep and Donald G. Nieman, 201–224. Athens: University of Georgia Press, 2001.

———. "African Americans and the Meaning of Freedom: Washington County, Texas, as a Case Study, 1865–1886." *Chicago-Kent Law Review* 70, no. 2 (1994): 541–582.

———. "Black Political Power and Criminal Justice: Washington County, Texas, 1868–1884." *Journal of Southern History* 55 (August 1989): 391–420.

Otto, John Solomon. "The Migration of the Southern Plain Folk: An Interdisciplinary Synthesis." *Journal of Southern History* 51 (May 1985): 183–200.

Owsley, Frank L. "The Pattern of Migration and Settlement on the Southern Frontier." *Journal of Southern History* 11 (May 1945): 147–176.

Porter, Kenneth W. "Negroes and Indians on the Texas Frontier, 1831–1876." *Journal of Negro History* 41 (July 1956): 185–214.

Robinson, Willard B. "Early Anglo-American Church Architecture in Texas." *Southwestern Historical Quarterly* 94 (October 1990): 261–297.

Rothstein, Morton. "The Antebellum South as a Dual Economy." *Agricultural History* 41 (October 1967): 373–382.

Ruggles, Steven. "The Transformation of American Family Structure." *American Historical Review* 99 (February 1994): 103–128.

Schlumbohm, Jürgen. "From Peasant Society to Class Society: Some Aspects of Family and Class in a Northwest German Protoindustrial Parish, 17th–19th Centuries." *Journal of Family History* 17 (April 1992): 183–199.

———. "The Land-Family Bond in Peasant Practice and in Middle-Class Ideology: Evidence from the North-West German Parish of Belm, 1650–1860." *Central European History* 27, no. 1 (1994): 461–477.

Schmitz, Mark. "The Transformation of the Southern Cane Sugar Sector, 1860–1930." *Agricultural History* 53 (January 1979): 270–271.

Scott, Rebecca J. "Defining the Boundaries of Freedom in the World of Cane: Cuba, Brazil, and Louisiana after Emancipation." *American Historical Review* 99 (February 1994): 70–102.

———. "Small-Scale Dynamics of Large-Scale Processes." *American Historical Review* 105 (April 2000): 472–479.

Shlomowitz, Ralph. "'Bound' or 'Free'? Black Labor in Cotton and Sugarcane Farming, 1865–1880." *Journal of Southern History* 50 (November 1984): 569–596.

Shook, Robert W. "The Texas 'Election Outrage' of 1886." *East Texas Historical Journal* 10 (Spring 1972): 20–30.

Sitterson, J. Carlyle. "The Transition from Slave to Free Economy on the William J. Minor Plantations." *Agricultural History* 17 (October 1943): 216–224.

Sparks, Randy J. "John P. Osterhout: Yankee, Rebel, Republican." *Southwestern Historical Quarterly* 90 (October 1986): 111–138.

Thornton, John. "Central African Names and African-American Naming Patterns." *William and Mary Quarterly*, 3rd ser., 50 (October 1993): 727–742.

———. "Religious and Ceremonial Life in the Kongo and Mbundu Areas, 1500–1750." In Heywood, *Central Africans and Cultural Transformations*, 72–90.

Tyler, Ronnie C. "Cotton on the Border." *Southwestern Historical Quarterly* 63 (April 1970): 456–477.

———. "Fugitive Slaves in Mexico." *Journal of Negro History* 57 (January 1972): 1–12.

Von Hinueber, Caroline. "Life of German Pioneers in Early Texas." *TSHAQ* 2 (January 1899): 227–232.

Weber, David J. "The Spanish Borderlands of North America: A Historiography." *Organization of American Historians Magazine of History* 14 (Summer 2000): 5–11.

Wolfe, Patrick. "Land, Labor, and Difference: Elementary Structures of Race." *American Historical Review* 106 (June 2001): 866–905.

Wunder, John R., and Pekka Hämäläinen. "Of Lethal Places and Lethal Essays." *American Historical Review* 104 (October 1999): 1229–1234.

Dissertations and Theses

Allison, Randall Scott. "A Dialogue between Material Remains, Historical Documents, and Oral History: Allenfarm and Rogers Plantation, a Case Study." Ph.D. diss., Texas A&M University, 1996.

Borchardt, Craig William. "German Lutheran Transplants and Methodist Converts in Washington and Austin Counties, Texas, 1860–1930." Ph.D. diss., Texas A&M University, 1996.

Curlee, Abigail. "A Study of Texas Slave Plantations, 1822–1865." Ph.D. diss., University of Texas, 1932.

Harper, Cecil, Jr. "Farming Someone Else's Land: Farm Tenancy in the Texas Brazos River Valley, 1850–1880." Ph.D. diss., University of North Texas, 1988.

Kelley, Sean M. "Plantation Frontiers: Race, Ethnicity, and Family along the Brazos River of Texas, 1821–1886." Ph.D. diss., University of Texas, 2000.

Kite, Jodella D. "A Social History of the Anglo-American Colonies in Mexican Texas, 1821–1835." Ph.D. diss., Texas Tech University, 1990.

Kuffner, Cornelia. "Texas Germans' Attitudes toward Slavery: Biedermeier Sentiments and Class-Consciousness in Austin, Colorado, and Fayette Counties." Master's thesis, University of Houston, 1994,

Platter, Allen Andrew. "Education, Social, and Economic Characteristics of the Plantation Culture of Brazoria County, Texas." Ph.D. diss., University of Houston, 1961.

Powers, Betty Jane. "From Cotton Fields to Oil Fields: Economic Development in a New South Community." Ph.D. diss., University of Houston, 1994.

Robbins, Fred. "The Origins and Development of the African Slave Trade into Texas, 1816–1860." Master's thesis, University of Houston, 1972.

Robinson, Armstead. "Day of Jubilo: Civil War and the Demise of Slavery in the Mississippi Valley, 1861–1865." Ph.D. diss., University of Rochester, 1976.

Summers, Suzanne Lynn. "The Role of Merchants in Promoting Economic Development in the Houston-Galveston Region, 1836–1860." Ph.D. diss., University of Texas, 1996.

Ward, Forrest Elmer. "The Lower Brazos Region of Texas, 1820–1845." Ph.D. diss., University of Texas, 1962.

Index

Index